Measuring and Evaluating Sustainability

The indexes used by local, national, and international governments to monitor progress toward sustainability do not adequately align with their ethical priorities and have a limited ability to monitor and promote sustainability. This book gives a theoretical and practical demonstration of how ethics and technical considerations can aid the development of sustainability indexes to overcome this division in the literature and aid sustainability initiatives.

Measuring and Evaluating Sustainability develops and illustrates methods of linking technical and normative concerns during the development of sustainability indexes. Specifically, guidelines for index development are combined with a pragmatic theory of ethics that enables ethical collaboration among people of diverse ethical systems. Using the resulting method of index development, the book takes a unique applied turn as it ethically evaluates multiple sustainability indexes developed and used by the European Commission, researchers, and local communities and suggests ways to improve the indexes. The book emphasizes justice as it is the most prevalent ethical principle in the sustainability literature and most neglected in index development.

This volume is an invaluable resource for students, researchers, and professionals working on sustainability indicators and sustainability policy-making as well as interdisciplinary areas including environmental ethics, environmental philosophy, environmental or social justice, ecological economics, businesses sustainability programs, international development, and environmental policy-making.

Sarah E. Fredericks is an Assistant Professor in the Department of Philosophy and Religion Studies at the University of North Texas, USA.

Routledge Studies in Sustainable Development

Institutional and Social Innovation for Sustainable Urban Development
Edited by Harald A. Mieg and Klaus Töpfer

The Sustainable University
Progress and prospects
Edited by Stephen Sterling, Larch Maxey and Heather Luna

Sustainable Development in Amazonia
Paradise in the making
Kei Otsuki

Measuring and Evaluating Sustainability
Ethics in sustainability indexes
Sarah E. Fredericks

Measuring and Evaluating Sustainability

Ethics in sustainability indexes

Sarah E. Fredericks

First published 2014
by Routledge
2 Park Square, Milton Park, Abingdon, Oxon OX14 4RN

Simultaneously published in the USA and Canada
by Routledge
711 Third Avenue, New York, NY 10017

Routledge is an imprint of the Taylor & Francis Group, an informa business

© 2014 Sarah E. Fredericks

The right of Sarah E. Fredericks to be identified as author of this work has been asserted by her in accordance with sections 77 and 78 of the Copyright, Designs and Patents Act 1988.

All rights reserved. No part of this book may be reprinted or reproduced or utilised in any form or by any electronic, mechanical, or other means, now known or hereafter invented, including photocopying and recording, or in any information storage or retrieval system, without permission in writing from the publishers.

Trademark notice: Product or corporate names may be trademarks or registered trademarks, and are used only for identification and explanation without intent to infringe.

British Library Cataloguing in Publication Data
A catalogue record for this book is available from the British Library

Library of Congress Cataloging-in-Publication Data
Fredericks, Sarah E.
Measuring and evaluating sustainability : ethics in sustainability indexes / Sarah E. Fredericks.
pages cm. -- (Routledge studies in sustainable development ; 4)
Includes bibliographical references and indexes.
1. Sustainable development. 2. Sustainability--Measurement.
3. Sustainability--Indexes. 4. Sustainability--Moral and ethical aspects.
5. Environmental ethics. I. Title.
HC79.E5F6957 2013
174'.4--dc23
2013002713

ISBN13: 978-0-415-83637-1 (hbk)
ISBN13: 978-0-203-69431-2 (ebk)

Typeset in Goudy and Times
by Taylor & Francis Books

Printed and bound in the United States of America by Publishers Graphics, LLC on sustainably sourced paper.

To Penny and Eldon Bryant
and Sylvia and Dave Fredericks
for your continual support of all I do.

Contents

Acknowledgments ix

1 **Introduction** 1
   1.1 Indexes and ethics 2
   1.2 Chapter overview 9
   1.3 A note on terminology 14

2 **Sustainability: a technical and normative endeavor** 15
   2.1 Precursors to the sustainability movement 16
      2.1.1 History 16
      2.1.2 Definitions 19
   2.2 The sustainability movement emerges 21
      2.2.1 The Brundtland Report 21
      2.2.2 For the Common Good 24
      2.2.3 Rio Earth Summit 27
   2.3 Diversification after Rio 36
      2.3.1 Sectorial definitions 37
      2.3.2 Monitoring and reporting 39
      2.3.3 Ethical analyses 42
      2.3.4 The environmental justice movement 46
      2.3.5 Rio+20 51
   2.4 Implications 56

3 **Index theory** 58
   3.1 Definitions of indexes and indicators 58
   3.2 Frameworks for sustainability indexes 60
   3.3 Balancing polarities in index development 63
   3.4 Normative priorities and index development 66
   3.5 Incorporating ethics into sustainability indexes 69

4 **Comparative ethics for sustainability** 72
   4.1 Theory of pragmatic ethical principles 75
      4.1.1 Peirce, Neville, and vagueness 75

viii Contents

      4.1.2 Muelder and middle axioms 79
      4.1.3 Broad principles for ethics: combining insights of vague categories and middle axioms 82
  4.2 *Ethical systems* 84
      4.2.1 Introduction to the ethical systems 85
      4.2.2 From comparisons to broad principles 96
      4.2.3 Broad principles of sustainability ethics – a working list 112
  4.3 *Potential critiques revisited* 116
      4.3.1 Why not the Earth Charter? 116
      4.3.2 Will the broad principles be acceptable? 117
  4.4 *Implementing the broad principles: a preliminary method* 118

5 An ethical examination of sustainability indexes 122
  5.1 *Carbon emissions indicators* 124
  5.2 *The three-dimensional index of sustainable energy development (SED)* 127
      5.2.1 Foundations of the three-dimensional index of SED 128
      5.2.2 The environmental dimension 130
      5.2.3 The social dimension 134
      5.2.4 The economic dimension 139
      5.2.5 An assessment of the entire three-dimensional index 146
  5.3 *Prescott-Allen's* The Wellbeing of Nations 148
  5.4 *The 2012 Environmental Performance Index (EPI)* 155
  5.5 *Eurostat's Sustainable Development Indicators (SDI)* 159
  5.6 *Local indexes* 164
  5.7 *Ethical strengths and weaknesses of sustainability indexes* 169

6 Environmental justice: a resource for sustainability indexes? 171
  6.1 *Definitions of environmental justice, their existing connections to and potential contributions to sustainability studies* 172
  6.2 *Methods of monitoring environmental justice* 174
  6.3 *Contributions of environmental justice studies to sustainability indexes* 179

7 Aggregating local indicators 184
  7.1 *Numerical aggregation methods* 184
  7.2 *Map-based aggregation methods* 188
  7.3 *Advantages of numerical and map-based aggregation methods* 192

8 Ethics in sustainability indexes 195

References 200
Index 219

Acknowledgments

While conceiving of and working on this project many people have assisted me, though of course any errors or omissions are my own. As this project arose out of my dissertation research, I must thank Binna Davidsdottir for introducing me to the idea of indexes; Wesley Wildman and Norm Faramelli for careful feedback and conversations about ethics and religion; and Robert Cummings Neville, Merlin Swartz, Cutler Cleveland, and Kirk Wegter-McNelly for their key comments and questions. Thanks to all of my colleagues at the University of North Texas who have discussed the project with me, advised me on publishing, and in general, encouraged me: Adam Briggle, J. Baird Callicott, Robert Figueroa, Robert Frodeman, P. (Trish) Glazebrook, Gene Hargrove, J. Britt Holbrook, Pankaj Jain, George James, David Kaplan, Irene Klaver, Ricardo Rozzi, and Martin Yaffe. I also greatly appreciate the fact that the Department of Philosophy and Religion Studies, the College of Arts and Sciences, and the University of North Texas encourages and enables such interdisciplinary work. Thank you also to my research assistants Sarah Conrad, Matthew Bower, Nick Sarratt, and Fábio Valenti Possamai. I am also grateful to financial support from the Department of Philosophy and Religion Studies at the University of North Texas for my research assistants and the work of Matt Story who helped prepare the index; the University of North Texas for my 2011 Junior Faculty Summer Research Fellowship, "Ethics in Sustainability Indexes," and 2009 Research Initiation Grant, "Monitoring Environmental Justice"; and the Center for the Study of Interdisciplinarity at the University of North Texas for my 2010 grant, "Philosophy and Monitoring Progress toward Sustainability." A special thanks to Khanam Virjee, Helen Bell, and Siobhán Greaney of Routledge for enthusiastically supporting this project and helping me with the technical details. I also appreciate the comments of the three anonymous reviewers of the manuscript.

Thanks to my friends and colleagues Aubrey Coleman, Julie Sorrell, Susan Engelhardt, Hee Kyung Kim, Derek Michaud, Catherine Klancer, Regina Walton, Kelly Wisecup, Judith Enriquez, Gaytri Mehta, Angelika Neudecker, Chris Hutchison-Jones, and especially Cristine Hutchison-Jones. They all have listened to me talk about indicators and ethics (maybe for longer than they would like), asked pertinent questions, and, at critical points, helped me take a break. I also greatly appreciate the support of my family, especially my parents Dave and Sylvia Fredericks; my sister, Rachel Fredericks; my grandparents Penny and Eldon Bryant and the late Earl and Lucille Fredericks; and my aunts and uncles, particularly Curt Bryant, Stephanie Beal, John Bryant, and Julie Cotter.

Thank you to the following publishers who graciously allowed me to revise and expand some of my previously published articles for inclusion in this book. Portions of Chapter 2 Section 2.3 were adapted from the author's article, Fredericks, S. E. (2012). Agenda 21. In W. Jenkins and W. Bauman (eds.), *Berkshire Encyclopedia of Sustainability* (Vol. 6: Measurements, Indicators, and Research Methods for Sustainability). Great Barrington, MA: Berkshire Publishing Co. Adapted with permission from Berkshire Publishing Group. Chapter 3 sections 3 and 4 were adapted from the author's article, Fredericks, S. E. (2012). Challenges to Measuring Sustainability. In I. Spellerberg, D. Fogel, S. E. Fredericks, L. M. Butler Harrington, M. Pronto and P. Wouters (eds.), *Berkshire Encyclopedia of Sustainability* (Vol. 6: Measurements, Indicators, and Research Methods for Sustainability). Great Barrington, MA: Berkshire Publishing Co. Adapted with permission from Berkshire Publishing Group. Chapter 5 sections 4–7 are an expansion of Fredericks, S. E. (2012). Justice in Sustainability Indexes and Indicators. *International Journal of Sustainable Development & World Ecology*, 19(6), 490–499, reprinted by permission of Taylor & Francis (http://www.tandfonline.com). Chapter 6, sections 2 and 3 on monitoring environmental justice are a revised and expanded version of Fredericks, S. E. (2011). Monitoring Environmental Justice. *Environmental Justice*, 4(1), 63–69. This material is reprinted with permission from ENVIRONMENTAL JUSTICE, Volume 4, Number 1, published by Mary Ann Liebert, Inc., New Rochelle, NY.

1 Introduction

Since the late 1980s "sustainability" and the related term "sustainable development" have grown from relative obscurity to popular ways of expressing the interconnection of environmental, economic, and social goals. Indeed, these ideas have been key parts of international, national, regional, and local governmental policies; business plans; mission statements of nongovernmental organizations including religious groups; and the ideals of average citizens. While there are many definitions of sustainability and sustainable development, let us look to the most common for now, "meet[ing] the needs of the present without compromising the ability of future generations to meet their own needs" (World Commission on Environment and Development 1987: 8). Imagine that citizens of a nation have this vision of sustainable development as one of their primary goals. To move towards it they begin making laws: to combat environmental degradation they mandate composting and recycling and regulate air pollutants and nuclear waste. Since they maintain that sustainability requires economic stability they devise incentives for green businesses. They also believe that a safe environment and strong economy will not mean much to humans if they do not have a high quality of life so they encourage health care, meaningful work, and the enrichment and preservation of cultural traditions for all people through a variety of policies and programs. But as is typically the case, policy initiatives alone will not be satisfying to the people; they will also want to know whether their new initiatives actually aid movement toward sustainability. Are the air and water cleaner than they once were? Is the economy able to thrive within environmental restrictions? Are all people able to live healthy and fulfilling lives? Will all people, segments of the economy, or ecosystems benefit from the new policies? If not, what factors determine uneven distribution of benefits and burdens? To answer these questions, they will turn to indicators and indexes, tools, usually quantitative, used to monitor progress toward a goal.

Yet as Chapter 5 will show, sustainability indexes only align with several of the most central ethical claims of the sustainability movement, including that of justice between and among generations, in the most cursory way. This mismatch matters because indexes drive social behavior: what they monitor often becomes *the* aim of future policy and action. Thus, the policy-makers, businesses, nongovernmental organizations, and citizens are unlikely to undertake activities that faithfully exemplify their vision of sustainability, let alone make progress toward it, if they use such indexes. Motivated by the need to alleviate such symptoms of disconnection to foster movement toward sustainability, this book develops a rigorous dialog between index theory and the environmental ethics of diverse ethical worldviews and demonstrates how such a method can improve current and future indexes.

1.1 Indexes and ethics

Indexes and indicators link an ideal vision of a system with a means of measuring progress toward it, often through mathematical functions which summarize complex information about the system. Indexes are comprised of many discrete bits of data, or indicators. Some indexes also employ multiple subindexes which focus on different components of the system, each of which may be comprised of many indicators. One common index is the grade point average found on a student's report card at the end of a semester. Rather than listing grades for many homework assignments, quizzes, papers, and exams (individual indicators), the report card summarizes the student's achievement for the semester through the grades earned in individual classes (subindexes) and with the student's semester grade point average (an index). Similarly, if one wonders how well a nation, city, or company is progressing toward sustainability, it would be cumbersome and confusing to list the results of every water quality sample, name every acre of land turned into a park, record every company's quarterly earnings, and track every health statistic of its people. Rather, one would desire an overall assessment of the nation's progress toward sustainability, maybe comprised of a few subindexes that track key aspects of sustainability such as its environmental, economic, and social dimensions. Indeed, since the late 1980s, thousands of sustainability indicators and indexes have been developed for nations, states, cities, companies, nongovernmental organizations, schools, and individual households (International Institute for Sustainable Development 2012, Krank and Wallbaum 2011: 1385, Spellerberg et al. 2012).

While indicators of one sort or another have existed for millennia (e.g. Nile height measurements were used as indicators of that year's agricultural success in ancient Egypt), their use has increased dramatically in recent decades. This prioritization of assessment is illustrated in American educational reforms such as the No Child Left Behind Act with its emphasis on standardized testing, in the international and national use of Gross Domestic Product (GDP) and the Human Development Index (HDI) as measurements of a country's economy and quality of life respectively, in calls for assessment of the outcomes of science research by granting agencies, and in a host of environmental assessments.

Multiple trends contribute to the increases in index popularity. As people recognize the complex dynamic nature of the interaction between humans and the rest of ecosystems, they desire methods of understanding and summarizing these relationships. New data collection methods, aggregation techniques, and storage capacities fueled by the computer industry have made it possible to construct much more complex indexes. The modern emphasis on reductionism has led people to believe that monitoring small pieces of larger systems can yield valuable information about the whole. Simultaneously, quantitative results are typically deemed more valuable, trustworthy, or real than qualitative sources. Thus, society expects and often requires quantitative indicators to demonstrate that there is a problem or whether progress has been made toward a solution regarding ecosystems, economies, individual health, or societal functioning. Democracies also play a role in the process insofar as taxpayers and voters want to know the efficacy of programs touted by politicians and supported by their tax dollars. This desire for knowledge has spread beyond the political sphere as consumers desire information about the environmental or labor practices of a company or the energy efficiency of its products (e.g. Energy Star labels). Shareholders also desire indicators of company performance, and increasingly, sustainability.

Indeed, indexes and indicators are not merely data organizers. If used, they drive feedback loops of social learning, decision-making, and action. For example, when indexes indicate that existing actions, whether study habits or environmental policies, support one's goals, people generally continue or increase such activities. Alternatively, if students' grades indicate that they did not master the material, they may analyze what they did well and use new study strategies in the future. Similarly, if indicators reveal that pollution levels in a local lake are staying constant or rising over time, a community may enact new pollution regulations or ensure that existing laws are enforced. In all of these cases, people act to raise the index or indicator score in order to move toward their goals, whether of education or sustainability.

Indexes and indicators can, however, drive a community away from its goals if the goals and indexes are not well aligned. For instance, picture a community that values critical thinking and analytical writing but only grades students on definitions and basic facts because such exams are easier to grade. If grades are emphasized in such a community, students and teachers will probably begin focusing their time on memorization drills and test taking strategies rather than on critical thinking and writing skills. Similarly, GDP is often used as a measure of a country's economic strength, and in part because its results are so widely available, is frequently taken as a sign of the overall well-being of a nation. Yet, since GDP is tied to formal markets, it does not track many critical factors for a country's well-being including natural resources or human health. Consequently, GDP may rise if a nation's ecosystem is destroyed for a narrowly measured short-term financial gain or if people fall ill and receive costly medical treatment rather than cheaper preventative care. If GDP is taken as *the* measure of a country's well-being, preserving natural resources or encouraging preventative health care will not be prioritized even though doing so may facilitate well-being as envisioned by the nation's citizens. In these cases, indexes can drive community actions away from reaching their goals even as index results may fool their users into thinking they are progressing toward their goals.

Thus, if anyone, whether a scientist, policy-maker, index developer, business leader, nonprofit organizer, ethicist, or average citizen wishes to promote sustainability paying attention to indexes and indicators is a strategic move. By doing so, they can help ensure that indexes drive public policy, corporate strategies, and the actions of nongovernmental organizations in desirable ways, in this case toward sustainability.

To ensure that indexes, whether grades or sustainability indexes, support their user's goals, multiple levels of evaluation and adaptation are needed. First, as an index is proposed and developed, attention should be paid to what it indicates about the state of the system in question, how it will push the actions or policies of the country, and whether it actually aligns with the goals of the community in question. Evaluation along all of these lines should continue during and after its implementation. Continual, or at least periodic, assessment of indexes and their impact is particularly critical when monitoring complex dynamic systems. For instance, attempts to monitor sustainability face the possibility that knowledge of and ideals for ecosystems, economic systems, society, and their interaction will change over time even if physical conditions remain the same. Of course it is likely that physical conditions of ecosystems and societies will also change, so indexes must be able to be adapted to new physical and cultural situations if they are to encourage progress toward the goals of a community, business, or nation.

This book illustrates such evaluation by analyzing whether and to what degree existing sustainability indexes align with the goals of their users and then suggesting ways to resolve any misalignment. In general, I observe that there is often a division between the goals of sustainability and the tools used to measure progress toward it. Consider, for instance, the WCED definition quoted above, which is often used to inform definitions of sustainability: "meet[ing] the needs of the present without compromising the ability of future generations to meet their own needs" (World Commission on Environment and Development 1987: 8). In the elaboration of this definition in the rest of the Brundtland Report, it is clear that the writers of the report prioritized a form of justice as they prioritized equitable access to basic goods and services necessary for life among people alive today and between this and future generations. This ethical commitment is echoed throughout the sustainability movement. Yet, as Chapter 5 demonstrates, indexes only align with central ethical claims of the sustainability movement, such as those regarding justice, in a cursory manner. Indeed, index developers often presume that moving a society, country, or the planet as a whole toward sustainability will necessarily ensure that all of its individuals also move toward this goal, what I call trickle-down justice. In other words, if people decrease pollution, increase recycling, increase the use of renewable energy, and so on then such efforts will help everyone today and in the future, because "all are connected." With this presumption, ethical claims of equity or justice may seem to be background assumptions of any index and thus do not need to be explicitly included as indicators in an index. The vast literature on environmental injustice, however, calls this line of thinking into question as it observes that subcommunities around the world, whether racially, ethnically, or economically separated from "average" people may have much worse environmental and social situations. These results cannot merely be explained by market forces as people of color are disproportionally affected at rates higher than one would expect when controlling for income or economic class (Agyeman et al. 2003, Bullard 1993). Indeed, environmental justice research suggests that sustainability initiatives do not ensure that the lot of all people will be improved unless specifically monitored and acted toward. Consequently, existing sustainability indexes will not intentionally drive their users toward justice, and therefore sustainability, as their users desire.

The misalignment between indexes and justice is an example of the much larger and widespread emphasis on the technical aspects of sustainability, what can be sustained in a given set of circumstances, to the degree that normative visions of sustainability, what people want to sustain, are overlooked. While the focus on technical assessments occurs through the sustainability movement, it is particularly acute in the development and use of indexes, as demonstrated in Chapter 2. When sustainability indexes or indicators do align with principles of sustainability ethics, as when they favor efficient and frugal use of natural resources, it is due to the implicit role of ethics in sustainability. Such alignment cannot be guaranteed with current theories of index development that rarely, let alone sufficiently, consider or advocate for the role of ethics in index development.

Indeed, there is a general lack of attention to the role for ethics in index development by index developers and ethicists on their own and in collaboration. When the two groups have interacted, the results have been insufficient to address the pertinent questions or significantly influence the field of index development. Admittedly, some index developers do recognize that normative priorities play a role in index development (Bleicher and Gross 2010: 603, Burger et al. 2010, Olalla-Tárraga 2006, Walter and Stuetzel 2009) and a few even recognize the need to consider diverse ethical perspectives to ensure that the indexes are ethically sound before they are adopted and more difficult to revise (Dahl 1997: 82), but this work has not yet been done. The few ethically minded scholars who have worked in the area have produced good work, but it is often cursory, abstract, and/or in need of updating given recent developments in the field (Daly et al. 1989, Norton 2005, Peet and Bossel 2000). For example, Herman Daly and John B. Cobb Jr's

1989 *For the Common Good* was a groundbreaking expansion of economic indexes to include social and environmental aspects of sustainability for a host of technical and philosophical reasons (Daly et al. 1989). Yet since then research regarding social and ethical aspects of sustainability has advanced significantly, particularly regarding justice. Additionally, Daly and Cobb's reliance on their own ethical position insufficiently addresses the challenges of ethical diversity in cross-cultural sustainability efforts. The most recent step toward representing the ethical in indicators or indexes is the movement to include diverse stakeholder participation in indicator formation, which presumes that involving diverse participants will enable their values to be included in the index (Elling 2008, Fraser et al. 2006, Lopez-Ridaura et al. 2002, Norton 2005). Bryan G. Norton's work is the most prominent in this genre. He develops a theory of collaborative, pragmatic decision-making that focuses index developers on particular situations rather than on ultimate ideals to avoid bogging conversations down in endless discussions of metaphysical ideals (Norton 2005). His approach, however, may not pay enough attention to the religious and philosophical foundations of particular groups' ethical commitments for their own satisfaction. Furthermore, his emphasis on theory and his commitment to users developing their own indexes means that he does not analyze specific indexes. Yet, examples of the ethical strengths and weaknesses of particular indexes are critical for improving existing indexes and illustrating the process and benefits of including ethics in index development, especially for people who are not trained ethicists or are ethicists unfamiliar with indexes.

Collaboration between index developers and ethicists is rare for at least four groups of reasons. First, there is a widespread assumption in modern Western society that technical and ethical assessments are, and should be, completely separable. Under the most naive versions of this view, scientific observations and experiments determine the way the world works and the best way it can function (proponents of this view do not recognize that "best" is a normative claim). A slightly more realistic view holds that scientific observations and experiments determine what is. Then communities, guided by ethical and cultural values can decide which of the possible alternatives they desire and ask the scientists (natural or social) how to get to that goal. In neither of these views is it necessary for natural and social scientists to consider community values when assessing the state of the world or compiling it into indexes because that data is "objective." According to these views, sustainability studies and indexes primarily do and should regard what can be technically sustained in an ecosystem – how many animals can be killed for food, how many crops grown, or how much waste assimilated while sustaining the ecosystem.

Yet a significant body of literature in the philosophy and sociology of science maintains (and supports with case studies) that facts and values are not completely distinct entities. Rather, one's own ethical values; the values, assumptions, terminology, and methods of one's epistemic community; and physical conditions shape one's scientific and technical assessments of the world. At least three broad types of interaction between facts and values exist: 1) in the identification of and the selection of problems for study, 2) as culture and terminology shape the way phenomena are understood, and 3) as norms influence thresholds of acceptable risk or change. These assertions do not mean that knowledge is *only* a social construction, for the world can also push back on our theories, but pure observation alone does not determine technical assessments and data. In relationship to sustainability this means that our values shape our understanding of the world and our assessments of what we want to sustain, for whom and for how long. Thus, calls to focus sustainability analysis and indexes on pure facts will fail because they will be impossible to construct without some interaction with values.

The historical and contemporary use of the term sustainability supports these broad theories of the interaction between facts and values. While this history will be discussed at length in Chapter 2, a brief example here will clarify this point. Recall the Brundtland definition of sustainable development discussed above. Note that its focus on justice is a normative, or ethical, claim for sustainability. Indeed, it is one of the most common ethical commitments in the sustainability literature. The writers of the Brundtland Report could have argued that sustainability involves preserving a good human life in symbiosis with the nonhuman ecosystem as it was at some point in the past and that the human population should be limited to 1/2 or 1/3 or 1/10 of the present population in order to make that happen. Such a decision, given the growing human population and our capacity to modify our environment, would probably make it technically easier to move toward sustainability. But that was not ethically acceptable to the authors of the Brundtland Report and the thousands of individuals and organizations who are influenced by this definition. Similar emphases on intra and intergenerational justice and a number of other norms including responsibility, careful use of resources, attention to the specific situation, adaptability, and far-sightedness can be found throughout the sustainability movement from its earliest to its latest formulations, from its most technical to its most explicitly ethical. Thus arguments for the separation of facts and values in sustainability studies do not reflect how the sustainability movement actually operates.

A second reason that ethics are rarely discussed in relation to sustainability indexes is that even if people recognize the value of explicitly integrating ethics into indexes, they may not have the tools or training to do so. Index developers are usually experts in economics, other social sciences, or environmental assessment and typically do not have the training to recognize when they are using ethical language or to consider the values of those impacted by the index when constructing it. On the other hand, ethicists are rarely trained in the various disciplines contributing to index theory. Additionally, both groups may also be wary of the others' fields and may experience disciplinary pressure to keep their distance.

Certainly, disciplinary divides can be difficult to overcome, but there are some scholars who recognize the need for such work and a few who have begun such endeavors, as noted above. Now, before insufficient indexes are even more firmly entrenched in policy-making and public awareness, is the time to step toward overcoming such divisions by critiquing, combining, and extending such work.

Discipline-specific issues also make index development difficult. For instance, indexes always simplify systems and thus may ignore key components or obscure crucial assumptions from average citizens. The challenge of identifying and aggregating indicators that adequately represent narrowly defined technical aspects of sustainability is only complexified by political, technical, monetary, and temporal limits to data availability. Quantitative methods of assessing the contribution of any one component of an ecosystem, say a species, to the functioning of the whole are hotly debated in ecology, yet such methods are often critical to environmental and sustainability indexes. Furthermore, quantitative measurements may not, and may never fully capture the depth and intricacies of ethical, aesthetic, recreational, and cultural values of ecosystems, particular species, one's home, or sacred land, or the intrinsic value of biota, land, and ecosystems, if such values exist. Thus even if index developers desire an ethically nuanced index they may not have the data or analytic tools to do so.

While such challenges cannot be fully resolved, attention to existing frameworks and guidelines as in Chapters 3 and 7, and the recognition that indexes and indicators can be useful but are not perfect and should always be open to improvements to enable us to capitalize on the possibilities of indexes without reifying them.

Moving to ethics, one quickly realizes that in today's pluralistic world, discussions of ethics in public policy-making raise thorny questions of whose ethics to use: those of ethical experts, governmental leaders, the majority, a minority, or some combination of the above? A "secular" or "religious" ethic? Ethics developed specifically for contemporary environmental situations or more traditional ethics which may have a history of negative environmental considerations? Furthermore, who should get to decide which ethics to use? Answering these questions is necessary when incorporating ethics into sustainability indexes because the specific goals and values of communities and nations may differ. Even if everyone in a nation completely agreed about their ethical point of view, these questions would arise since cooperation between local endeavors will be necessary to make significant progress toward sustainability at a global level. The pragmatic ethical method outlined in Chapter 4 in which broad principles of sustainability ethics are identified through a rigorous process of comparison with multiple religious and philosophical systems answers this need. The principles identified through this method can be specified in various, even contradictory, ways by people of different ethical traditions, providing broad constraints for ethical decision-making in multi-ethical settings. Thus this method mediates between allegiance to a traditional ethical system and methods of decision-making which neutralize ethical differences by focusing on concrete problems.

While there are significant obstacles to incorporating ethical concerns into sustainability indexes, the resources are available to do so and it is a worthy goal if one aims to advance toward sustainability. After all, indexes can be quite valuable when developing societal goals and assessing progress toward these goals. Furthermore, indexes are so engrained in modern society and expected in policy-making that to refuse to engage in discussions about indexes and indicators because they are not adequate means of monitoring many types of values is to recuse oneself from pressing public-policy discussions about the future of human society on earth. On the other hand, developing indexes without explicit attention to ethics runs the risk of ineffective indexes, or even worse, indexes which drive people away from their vision of sustainability. A more effective strategy requires investigating how indexes in general, and sustainability indexes in particular, can better reflect technical considerations as well as the specific normative priorities of individual communities and the common ethical commitments of the international community, the aim of this book.

1.2 Chapter overview

To develop methods for incorporating ethics into sustainability indexes, an understanding of the way people envision sustainability, as well as a firm grasp on indicator theory and ethical theory, the subjects of the first three chapters, are necessary. The final four chapters apply these lessons to indexes of sustainability and sustainable development, noting where the indexes do and do not align with ethics of sustainability and building on literature about environmental justice to begin to fill this gap.

Chapter 2 encompasses three main sections. First a study of the antecedents of the sustainability movement and the multifaceted connotations of the word "sustain" demonstrates that both technical and ethical considerations played a significant role in the historical and linguistic foundations of the movement. Next, early documents of the movement including the Brundtland Report, Herman E. Daly and John B. Cobb Jr.'s *For the Common Good*, and the documents of the Earth Summit at Rio de Janeiro consider both technical and normative aspects of sustainability even though the technical are given priority and the two are increasingly separated (Daly et al. 1989, Robinson et al. 1993, United Nations Conference on Environment and Development 1992, World Commission on Environment and Development 1987). As the movement expanded and diversified in the 1990s, definitions of sustainability for particular sectors of society and methods of monitoring sustainability were developed. These technical studies were explicitly disassociated from normative investigations of sustainability as the sustainability movement increasingly developed along disciplinary lines even though normative assumptions implicitly influenced even the most technical studies. When ethical and technical evaluations seem most integrated, as in the environmental justice movement and the Rio+20 Conference, this integration fails to penetrate to the level of indexes. This bifurcation left many questions about sustainability, particularly about monitoring sustainability, unanswered and unanswerable, hindering understanding of and progress toward sustainability. To address these issues, a greater integration of technical and normative aspects of sustainability is necessary. Since the literature has treated technical and normative aspects of monitoring sustainability so separately, Chapters 3 and 4 explore these aspects of sustainability in greater detail before they are integrated in later chapters.

Chapter 3 outlines the reasons that indicators are used and the challenges in creating and using them. The modern preference for quantification and reductionism coupled with the ideal of transparency in policy-making that surfaces in democracies contributes to the popularity of indicators. Indeed, these assumptions are so strong that they may lead people to reify index results. When they are recognized as what they are – useful but limited tools for understanding complex dynamic systems – then they can considerably enhance our understanding of systems to be sustained. In particular indicators can monitor the driving forces upon, the state of, or the responses of elements in a complex, dynamic system (Mortensen 1997). These assessments may be used to evaluate the outcomes and cost-effectiveness of existing policies for nations, businesses or nongovernmental organizations and to guide future policy-making decisions.

When constructing indexes, index developers work to balance a number of factors. First, the desire to have technically rigorous, empirically sound indicators must be weighed against the need for indicators to be understandable by the average citizen or policy-maker. Second, the desire for indexes to comprehensively monitor all aspects of a complex system must be weighed against the desire for manageable indexes that are not too cumbersome, costly, or time-consuming to use. Third, index developers recognize that data that would ideally be used to monitor an index may not be available given current data collection methods and the realities of funding, time, and political constraints. When ideal data is not available, index developers must identify proxies to monitor the part of the system in question or choose not to monitor at all. Finally, as I will argue throughout this book, index developers must recognize the interaction between technical assessments and normative priorities. I suggest that index developers should do this by explicitly enabling ethical principles to guide index construction, use, and evaluation in general *and* by enabling ethical priorities to shape or suggest specific indicators for their indexes.

Recognizing that there is a need to explicitly involve ethics in the development of indicators, the first section of Chapter 4 develops a pragmatic method of identifying ethical principles for sustainability while the second employs this method to identify such principles. The method developed here combines the semiotic theory of Charles Saunders Peirce as developed by Robert Cummings Neville and Wesley Wildman for use in comparative religious studies with methods of comparative ethics in the work of Walter Muelder and the Earth Charter (Muelder 1959, Muelder 1966, Neville 2001a, Neville 2001b, Neville 2001c, The Earth Charter Commission 2000). The result is an iterative, adaptive, collaborative approach to ethics which articulates broad principles of sustainability ethics that resonate with the deeply held convictions of people from many different religious, philosophical, and cultural traditions and thus can facilitate collaborative decision-making while establishing general parameters of ethical acceptability. For instance, equitable access to resources necessary to meet one's needs is often articulated as a component of sustainability. Certainly it can be specified in a number of ways, as people define what exactly equity is, who it should apply to and when, and what needs are. Yet even without agreement on the exact application of equity in a particular case, a broad vision of equity can assess whether its general sense is represented in sustainability indexes.

This flexible approach enables a middle ground between extreme relativism, in which all ethical systems are deemed equally acceptable, and absolutism, in which one and only one ethical system is considered absolutely right for all people. As such, it answers the question of "whose ethics should be used in sustainability studies?" a critical question for scholars and lay people alike in our pluralistic, postmodern world. The principles also provide a cross-cultural language for ethical conversation as is needed during collaboration for sustainability on a community, national, or international level.

Additionally, since the method opens the principles to critique and corrections it limits the epistemological and social dangers of elevating ethical principles to the status of moral absolutes while avoiding the need to completely reinvent an ethical outlook each time one deliberates about ethical actions. Instead, the broad principles may be revised as more communities from different ethical systems enter the conversation, as ethical knowledge advances, as technical knowledge is gained, and as the environmental or social situation changes. Thus, the ethical principles are not absolute or ideal laws or rules, but are as open to revision as the technical and scientific understandings of the world they interact with. Now, this does not mean that the ethical principles will be revised every few minutes or even every few years, nor does it mean that the revised principles will have no connections to longstanding ethical traditions. Rather this theory of adaptable ethical principles explicitly enables ethical theorizing to continue as it has done for some time. Picture, for example, the changes to ethics that arose with the end of slavery and the rise of the civil rights movements. This iterative, collaborative ethical method is the best method of ethics in the absence of ethical certainty and in the presence of so many different ethical traditions.

In the second part of Chapter 4 I use the method developed in part one to identify a preliminary list of ethical principles for sustainability. All too often, academic ethicists ignore the ethical insights developed by people living out their everyday lives (cf. West 2006: 3–16). This trend means that professional ethicists not only exclude potentially valuable ethical insights from their analysis but also potentially alienate everyday people from their theories. Such a top-down model may demonstrate how to incorporate some ethics into sustainability indexes but will not enable the multifaceted approach needed to promote sustainability across cultures and at a range of scales. To counter these trends and recognize and build upon sustainability ethics articulated and endorsed by a vast number of people, acknowledging the normative insights of many people in addition to those of professional ethicists, I begin with the ethical principles embedded in Agenda 21, a widely endorsed and influential UN blueprint for sustainability. As many sustainability indexes follow the definition of sustainable development outlined in Agenda 21, starting with its ethics also enables the identification of ways in which indexes do not live up to its ethics.

Since Agenda 21 was not intended to be an ethical document its ethical principles are not always articulated in the detail that is necessary to use them to evaluate existing indexes and guide the development of new indexes. Thus I refine and elaborate upon the ethical principles of Agenda 21 (justice, responsibility, an adequate assessment of the situation, careful use of resources, feasibility, and farsightedness) through comparison with ethical principles developed by environmental ethicists from Christian, Islamic, and deep ecological perspectives as well as by the Earth Charter, a vision of environmental ethics intended to be global in perspective. As per the method, I welcome conversation with and critique from these and other perspectives though I selected a few for this project to keep its scope manageable.

Drawing on the investigations of sustainability, indicators, and ethics from previous chapters, in Chapter 5 I analyze the degree to which the ethical priorities for sustainability are incorporated into various sustainability indexes. Specifically, I assess the Environmental Performance Index, the Well-being of Nations, the Three-Dimensional Index of Sustainable Energy Development, carbon emissions indicators, Eurostat's Sustainable Development Indicators, and several local sustainability indexes (Davidsdottir et al. 2007, Eurostat 2012d, Fraser et al. 2006, Gallego Carrera and Mack 2010, Lopez-Ridaura et al. 2002, McMahon 2002, Praneetvatakul et al. 2001, Prescott-Allen 2001, Reed et al. 2008, Wei et al. 2007, Yale Center for Environmental Law & Policy and Center for International Earth Science Information Network 2010). This selection of indexes ensures that many types of index will be examined including those applicable to various scales from the local to national, at various degrees of complexity, for both specific areas of sustainability (energy) and sustainability as a whole. These analyses reveal that the indexes generally follow many ethical principles including responsibility, careful use, and feasible idealism; follow adaptability in some respects and not others; and fall shortest with respect to adequate data assessment and justice. Indeed, they only monitor justice in a cursory way because they rarely track whether subpopulations in a nation or a community experience differential access to environmental benefits or exposure to environmental burdens. Of course, this limitation hinders the indexes' alignment with other principles including adequate assessment of the situation, adaptability, responsibility, and farsightedness since the principles are interrelated. Given the prominence of justice in definitions of sustainability, this omission is striking and problematic.

To begin to address the lack of attention to justice in indexes, I assess and critique methods of monitoring environmental justice in Chapter 6. Distributive environmental justice, the equitable distribution of environmental benefits and burdens, is nearly identical to what sustainability scholars imply by intra and intergenerational equity and is already monitored using quantitative means. In the first section of Chapter 6, I explore how definitions of environmental justice complement ideas of equity in sustainability discussions and challenge existing trends of leaving justice and equity out of indexes. In the second, I categorize the methods used to monitor environmental justice into two broad categories: the first includes statistical studies and geographic information systems (GIS) analyses, both of which use quantitative data and typically monitor distributive issues. The second class of methods includes surveys and interviews which may monitor quantitative and qualitative aspects of environmental justice. This group monitors participation in decision-making and access to information as well as perceptions of distributive justice.

While hundreds of studies have monitored the existence of environmental injustices, nearly all focus on one narrowly defined issue (e.g. the location of toxic waste dumps) or on the issues facing a particular community (e.g. the south side of Chicago); many are narrow and local. This focus enables researchers to observe injustices that may be hidden at larger geographic levels of analysis and fosters the participation of local people, a valued aspect of environmental justice. Yet monitoring particular, local cases of injustice cannot immediately fill the gaps in sustainability indicators since sustainability indicators typically focus on national rather than community issues and often aim for comprehensive rather than issue-specific assessments of sustainability (Fredericks 2011). Thus the general scope and terms of environmental justice research can significantly bolster sustainability research, but new methods are needed to incorporate these elements into sustainability indicators that will be useful for guiding policy at a national or state-wide level.

Recognizing that resources for incorporating ethical concerns such as equity and justice into sustainability indexes are still in their infancy, in Chapter 7 I propose numerical and map-based methods of aggregating local assessments into wide-scale indexes. These methods can aggregate community-based indicators, including those of environmental justice, into state or national level indexes such that some of the community specificity is preserved. In this way I show how technical aspects of sustainability and ethical assessments from many different ethical systems can simultaneously influence index development.

While monitoring justice and other community-based indicators of sustainability is a critical move to improve the alignment of sustainability indexes with ethical principles, this move alone will not ensure that they align with normative visions of sustainability because sustainability indexes have other ethical limitations. General reflections on the ways ethics can be a part of the whole index development process are needed if ethical aspects of sustainability, including and beyond justice, are to be seriously considered during index development. Thus, Chapter 8 identifies the ways ethics can and do contribute to index development. Naming and enabling these roles is necessary to ensure that ethics continues to have a place in sustainability indexes beyond the particular critiques made in this text and indeed, in indexes in general.

1.3 A note on terminology

"Sustainability" and "sustainable development" are terms that can have significantly different meanings but are intimately related in the sustainability movement. Sustainability is the ability for a set of conditions, an ecosystem, business, or society, to be maintained. Sometimes it is taken to mean static preservation and sometimes it implies a dynamic integration in which parts of the whole may change, but a sense of or function of the whole is preserved. At minimum, sustainability is normative because its supporters assume that the ability to sustain something is generally good and desirable. Sustainable development, on the other hand, is usually focused on economies or on societies as a whole and may be more explicit about justice and other normative claims than sustainability insofar as these definitions proscribe a favored type of development. There have been many attempts to clarify such definitions and determine whether or not it makes sense to apply them to complex, dynamic systems. Despite these differences, popular and scholarly literature often elides the two. Additionally, and importantly for this study, definitions of sustainability and sustainable development both weave technical and normative ideas together, though their indexes consistently focus on the technical to the exclusion of the ethical. I will argue that the same method can be used to improve the indexes of sustainability and sustainable development. For all of these reasons, it is appropriate and useful to speak of the sustainability movement as one focused on defining and making progress toward goals of sustaining, sustainability, and sustainable development, including the construction and application of indexes. From time to time in general discussions, I use "sustainability" as a shorthand for this constellation of terms.

2 Sustainability
A technical and normative endeavor[1]

Technical and normative aspects of sustainability are always intertwined. One's understanding of the way the world works and, therefore, what can be technically sustained influences what one thinks is right or desirable; one's normative vision influences how one understands the world. Technically possible means of moving toward sustainability will not be fully implemented unless acceptable to society; normative visions of sustainability that are not technically possible will not be achieved. This mutual influence significantly predates the existence of the sustainability movement and continues to the present day. Yet the majority of the sustainability movement has separated deep explicit studies of the normative aspects of sustainability from the technical studies which often overshadow them. This bifurcation leaves a number of questions about the articulation of sustainability goals and methods of monitoring progress toward them unasked and unanswerable: How can technical methods of monitoring sustainability be constructed to best align with ethical priorities? How can ethical positions from different worldviews influence sustainability goals and indexes without reducing these positions to the least common denominator or developing multiple indexes each only suitable to a narrow ethical vision or community, thwarting the international cooperation that will be needed for global sustainability? Neglecting these questions hinders movement toward sustainability since sustainability policies and indexes developed in their absence will not necessarily attend to the fundamental links between the ethical and technical nor to the ethical diversity in the contemporary world.

To support these claims, this chapter examines the history of the sustainability movement, considering three broad time periods. The first, before the mid-1980s, includes the precursors to the sustainability movement. A study of the antecedents of the movement and the multifaceted connotations of the word "sustain" in this section demonstrate that both technical and ethical considerations played a significant role in the foundations of the movement. The second phase, from the 1986 publication of *Our Common Future* to the 1992 Earth Summit in Rio de Janeiro, is characterized by explicit international attention to sustainability. Major documents of this period including the *Our Common Future* (the Brundtland Report), Herman E. Daly and John

B. Cobb Jr.'s *For the Common Good*, and the documents of the Earth Summit at Rio de Janeiro (the Rio Declaration and Agenda 21) consider technical and normative aspects of sustainability, though they give priority to the technical. The third, post-Rio period witnessed a tremendous growth in sustainability studies. Definitions of sustainability for particular sectors of human activity such as agriculture and energy use proliferated as did methods of monitoring sustainability for businesses, countries, and individuals. Yet these analyses did not follow Daly and Cobb's integrated approach to sustainability indexes. Instead, they were largely explicitly separated from normative studies of sustainability though even the most technical discussions were permeated by normative assumptions and assertions, the most common being a commitment to justice among currently living generations and between present and future generations. As this historical analysis will show, the compartmentalization of technical and normative sustainability studies ignores a number of key questions about sustainability indexes and thus inhibits progress toward sustainability.

2.1 Precursors to the sustainability movement

2.1.1 History

Most histories of sustainability begin in the 1970s, though a few begin in antiquity, the nineteenth century, or the early twentieth century (Du Pisani 2006: 83–85, Johnston 2010, LeVeness and Primeaux 2004, Mitcham 1995: 312–13, Ricketts 2010). Regardless of where they begin, however, these histories illustrate the mutual interactions of technical knowledge about the world and people's normative priorities for themselves, their society, and the world at large.

Histories that trace the roots of this movement back to antiquity typically argue that Christian ideas of a linear progress throughout history (rather than cyclical visions of history) were foundational to the idea of preserving the status of the present or past for future generations (Du Pisani 2006: 83–85, Mitcham 1995: 312–13). They argue that a linear vision of history enables a sense of significant progression or denigration over time without the possibility of renewal in the next cycle of history. Thus, if the past or present is deemed good, or at minimum, better, than an unknown future, preservation may be prioritized. Scholars advocating this view of history assume that a cyclical view of time would not lead to a focus on sustaining the conditions of any one point in time because of the possibility of repetition and new chances in cycles of history. In any case, these analyses indicate that technical ideas about the way the world worked, for example a linear progression of time, were tightly coupled with a normative vision of a better future, or at least one that was no worse than the present.

Examining the intellectual history of eighteenth and nineteenth century Europe, J. A. Du Pisani maintains that ideas of inevitable progress were coupled with observations from the era to promote antecedents to the sustainability movement. During this time, some scholars recognized that the human population was growing and that there was an unequal distribution of benefits from the industrial revolution. Thus, they reasoned, life in the future was going to be worse for an expanding group of people if no provisions were made for the future. Following a nascent idea of intergenerational justice, some scholars advocated methods of using resources (including timber and coal) at rates that would ensure that some would be left for future generations (Du Pisani 2006: 83–85). Similar combinations of observations about the world, predictions of the future, and normative claims including intergenerational justice would come to pervade the sustainability movement.

Nature writers of the nineteenth century such as Emerson and Thoreau, the work of preservationists and conservationists including John Muir and Gifford Pinochet, national parks advocates and members of new clubs including the Sierra Club (1892) and the Audubon Society (1905) are also named as antecedents of the sustainability movement. The actions of these individuals and groups demonstrate how they coupled observations of the natural world with normative visions of how humans should act in and towards it (Johnston 2010: 177). Environmental awareness grew significantly in the 1960s as influential texts including Rachel Carson's 1962 *Silent Spring* and Paul and Anne Ehrlich's 1968 *The Population Bomb* enabled every-day citizens to connect their own experience to larger societal and environmental trends and formulate explicit normative positions on environmental issues (Carson 2002, Ehrlich and Ehrlich 1968).

The field of development studies has also significantly contributed to the sustainability movement. Lucas F. Johnston notes that the postwar founding of the Bretton Woods Institutions including the International Monetary Fund, the World Bank, and the World Trade Organization as well as President Kennedy's focus on the 1960s as the decade of development were signals of the growing international focus on economic and social development in the mid-twentieth century. He maintains that these wide-scale initiatives tied to capitalist economic systems, western governments, and western science led many to question what form of development was economically and socially preferable, helping to pave the way for the concept of sustainable development (Johnston 2010: 177–78). Prompted by mounting evidence that humans were having a significant impact on the environment that would prove harmful to humanity, the UN held the United Nations Conference on the Human Environment in Stockholm in 1972, an event which connected economic, social, political, and environmental problems on a global level in a global arena for the first time. It particularly linked poverty and environmental degradation, suggesting through its declaration and action plan what humans should do to ensure that meaningful, productive human life could continue (Johnston 2010: 178, LeVeness and Primeaux 2004: 186, Moffatt 1992: 28–29, United Nations Conference on the Human

Environment 1972). This commitment to a just distribution of the basic goods and services required for life would continue throughout the sustainability movement.

Links between assessments of the state of the world and normative visions of the world are also found in texts about economics and natural resources. In 1972 *The Limits to Growth* was published by the Club of Rome. It maintained that if current rates of consumption and population growth continued humans would demand more resources than the world had, leading to significant environmental and social declines. *Limits to Growth* sparked significant debates about whether continued economic growth was possible or desirable (Du Pisani 2006: 90, Meadows et al. 1972, Mitcham 1995: 314–415).

One of the earliest uses of the term "sustainable" in a context that acknowledged the limits of nature as well as environmental and social concerns came in the work of the World Council of Churches (WCC), an ecumenical group of hundreds of member churches representing millions of people which works to promote unity amid Christian diversity through prayer, worship, study, and social activism (World Council of Churches). In the late 1960s and early 1970s one of the WCC's foci was the influence of technology on society. Recognizing the impact of widespread use of certain technologies on nature as a part of its ongoing reflection on what makes up and how to achieve a responsible society, a 1974 report "accepted the thesis of nature's limits and called for a society that is both just and sustainable" (Chapman 2000, as quoted in Johnston 2010: 180–81). The WCC also demonstrated its growing tendency to include environmental concerns in its work as it reformulated its decades-old vision of the "responsible society" to a "Just Participatory and Sustainable Society" (JPSS) for its 1975 assembly in Nairobi (Wogaman 1993: 266). Reports from WCC assemblies during this period make it clear that its language of JPSS was responding to emerging technical data about human impact on the environment and society by making normative claims about what should be done in the face of such possibilities (though they often did not examine environmental issues as much as they could have given this language). In other words, they interwove the technical and ethical in their discussions of sustainability (World Council of Churches. Central Committee 1975, World Council of Churches. Central Committee 1983).

In the late 1970s and early 1980s international conferences not explicitly about sustainability including the Brandt Commission and the Palme Conference nevertheless highlighted ideas that would become central to the sustainability movement (Johnston 2010: 178, Mitcham 1995: 316–17, Thompson 2010: 197–200). For instance, the Palme Conference recognized that direct threats to the security of a few nations could threaten global security. Similar ideas would be stressed in the sustainability movement as people recognized the ways in which the sustainability of an individual region or nation depended on others because air, water, biota, and economies do not stop moving at national boundaries. Similarly, the Independent Commission on International Development, also known as the Brandt Commission for Willy Brandt, its leader, was primarily framed as focusing on relationships between countries in the global North and South, particularly on the flow of goods and services between these regions, rather than on environmental issues. Yet analyzing the movement of resources and products is an inherently environmental issue, recognition of which has been key to the sustainability movement. The report also contributed to the sustainability movement as it discussed many critical topics for the movement including "population, food and agriculture, energy, renewable resources, trade, and disarmament" (Thompson 2010: 198). Additionally, it named a variety of rights and responsibilities held by nations with respect to their relationships with other countries and the environment. Thus, it advocated that there were ethical, not just economic, reasons to attend to North–South relations (Thompson 2010: 198–99).

These many predecessors of the sustainability movement, whether from political, financial, or religious leaders; scholars of economics or natural resources; professional environmentalists; or average citizens all influenced the sustainability movement through the coupling of technical assessments and normative priorities. Aside from the WCC, however, "sustainability" had not yet been significantly used as a term to link economic, environmental, and social concerns. As "sustainability," and "sustainable development" gained traction in the academic and policy-making worlds, definitions of these terms were influenced by the social, political, and academic trends previously mentioned and by the broader associations and connotations of the term "sustain" within the English speaking world.

2.1.2 Definitions

Exploring the common meanings of the word "sustain" and its variants such as "sustainability" reveals the depth of interaction of the technical and normative aspects of sustainability. Recognizing this interconnection suggests that it is intellectually and socially problematic to emphasize one at the expense of another as has often happened in the movement.

The *Oxford English Dictionary* identifies multiple definitions of "sustain," many of which pertain to this discussion. These definitions can be divided into two categories: The first, the technical group, suggests that "sustain" can mean merely preserving the existence of an entity, such as when a work of art is conserved or an organization is kept up. Additionally, it can entail maintaining *life* through physical nourishment or emotional support. The second, moral, category, includes definitions such as "To uphold the validity or rightness of; to support as valid, sound, correct, true or just," and properly only connotes the preservation of entities or actions that are inherently valid or valuable (OED 2011a, OED 2011b). Yet, when the term "sustain" is used in everyday conversations and formal policy documents these moral connotations are often combined with technical definitions to yield the assumption that whatever one sustains or wishes to sustain *should* be sustained.

This conjunction is at the heart of the modern sustainability movement. Yet, many logical flaws can arise from conflating the two definitions, leading some to think that only the technical or, rarely, only the normative aspects, should be studied. For instance, humans technically could have continued to use chlorofluorocarbons (CFCs) rather than implementing a program to discontinue their use through the Montreal Protocol. After all, the ability to manufacture and use CFCs would not be impacted by the hole in the ozone layer until human life in general was significantly affected. And yet, because of their ethical priorities, people decided that the long-term consequences of using CFCs were significant enough to warrant drastically reducing their use (Attfield and Wilkins 1994: 158).

Similarly, just because people want something to be sustained upon initial consideration does not mean that it should be or technically can be sustained or that, upon reflection, they actually would want to sustain it forever (Beckerman 1994: 193). Most people have had an experience whether a meal, holiday, or engaging conversation that was so enjoyable that they wanted it to continue. Yet, humans generally realize that continually experiencing even these most pleasurable moments is not actually desirable. Without the contrast of our everyday, mundane existence, the peak experience may not seem as good. Furthermore, without the typical we may not have the temporal, monetary, physical, or emotional resources to live out the peak event. Prolonging one experience such as a favorite meal indefinitely could have disastrous effects on one's pocketbook and health; continuously eating is not technically sustainable. Clearly then, any priority that one wishes to sustain should not necessarily be sustained and may not be technically able to be sustained.

Despite the fact that conflating technical and normative aspects of the term "sustain" can lead to logical problems, the sustainability movement, as shown below, has consistently linked the two as it tries to identify modes of living that will enable societies and ecosystems that are technically capable of being sustained and will be morally acceptable. Moreover, such links are necessary since methods of understanding and technically assessing the world are influenced by one's worldview and one's normative priorities while one's normative priorities and worldview are shaped by one's experiences of and understanding of the world. Figuring out how to balance these desires without conflating them or totally separating them has been one of the great challenges of sustainability studies. Typically, problems arise when people claim that a specific act or condition is capable of being sustained while assuming that it *should* be sustained without acknowledging that their presuppositions of value may not be shared by all. Indeed, this issue is at the heart of many of the critiques of the most famous definition of sustainable development, that of the Brundtland Report.

2.2 The sustainability movement emerges

In the mid-1980s to early 1990s the sustainability movement emerged as a distinct endeavor in the international policy-making and academic arenas. Major documents of this period including the Brundtland Report (*For the Common Good*) and Agenda 21 developed definitions of sustainability and blueprints for monitoring and moving toward it. These documents tended to overtly emphasize technical aspects of sustainability but were certainly also infused with normative priorities. As the movement grew, disciplinary perspectives began to separate the study of technical and ethical elements of sustainability even though both continued to be implicitly influential.

2.2.1 The Brundtland Report

Prompted by increasing environmental problems and local and international attention to some combination of societal well-being, the state of the economy, and environment, and influenced by the many connotations of "sustain," the United Nations formed the World Commission on Environment and Development (WCED). The WCED, also known as the Brundtland Commission for its leader Gro Brundtland of Norway, was charged with discerning long-term environmental and social priorities for the "international community" to respond to environmental problems and develop strategies for sustainable development (World Commission on Environment and Development 1987: ix). Its report, *Our Common Future* (the Brundtland Report), was published in 1987. Famously, the Commission defined sustainable development as "meet[ing] the needs of the present without compromising the ability of future generations to meet their own needs" (World Commission on Environment and Development 1987: 8). This definition would become the most cited in the multiple disciplines that study sustainability; its publication and the wide-scale international response it provoked marks the beginning of the sustainability movement proper. Like its predecessors and the root term "sustain," the Brundtland Commission would interweave technical assessments of the state of the environmental and economy with normative visions of how the world should be. Yet the Brundtland Report also shifted toward emphasizing the technical, economic, and environmental over the social and normative aspects of sustainability.

For example, Paul Thompson's comparison of the Brandt and Brundtland Reports finds that both focus on "virtually identical" topics including "population, food and agriculture, energy, renewable resources, trade, and disarmament" without making specific policy recommendations (Thompson 2010: 198). Yet Thompson notes that while rights and responsibilities are explicit foundations of the Brandt Report, these elements are rarely discussed in the Brundtland Report. Instead, the Brundtland Report arrives at ethical considerations such as equity and the need to cooperate across national borders from the unbounded nature of ecological systems; if ecological systems do not follow national boundaries it argues that neither should human responses to environmental problems (Thompson 2010: 199). Thompson correctly argues that the Brundtland Report does not use the language of rights and responsibilities and that it emphasizes its technical foundations, but this does not mean that it lacks normative claims. Indeed, the Report was clearly influenced by a variety of ethical commitments though they were emphasized less than its technical components.

To understand the implicit normative position of the Brundtland Report one must first understand how it explicitly presents its priorities. The Report as a whole elaborates on its definition of sustainable development while exploring the world's economic, environmental, and social relationships; the problems that the world community faces; and ways humans may alleviate these problems. Throughout the Report, its recommendations build upon existing treaties and encourage collaboration between nations 1) to report on the status of the environment and pollutants, 2) to share scientific and technical information about the environment and economy, and 3) to develop laws. The Brundtland Report usually includes general recommendations such as monitoring and decreasing pollution rather than articulating specific policy recommendations such as an acceptable level of carbon emissions. It focuses upon gathering scientific data and the policy-making process, in other words, technical aspects of sustainability.

Though the Brundtland Report focuses on outlining a technical vision of sustainable development, evidence located throughout the document demonstrates that ethical principles including responsibility, an adequate and realistic assessment of data, equality and justice, adaptability, farsightedness, and possibly careful use of resources guided its writers. Thus, normative commitments are significant aspects of the text even if they are not explicitly enumerated. For instance, the Report acknowledges the deep interconnectedness between all people, economies, and ecosystems, and strives to promote the common good of all people. It recognizes that "the relative neglect of economic and social justice within and amongst nations" is often a cause of "our inability to promote the common interest in sustainable development" (World Commission on Environment and Development 1987: 49). Thus, it encourages prioritizing the needs of the poor so that all humans have equitable opportunities to meet their needs (44–49). Building from these ideas and its claim that we humans are responsible for the consequences of our actions, the Report suggests that we humans are responsible for changing our behavior so that present and future generations have the ability to survive. As it investigates the ways in which we should modify our behavior, the Report prioritizes an adequate and realistic assessment of data (52, and Ch 4–9). I name this an ethical principle because attending to the reality of the situation as described by scientists (modern and traditional) and experts of cultural analysis is necessary for an ethical response to particular ethical situations and advocating this is a normative claim. Acknowledging the dynamic nature of the world as described by the sciences, the Brundtland Report promotes the principle of adaptability. For example, it claims that sustainable development "is not a fixed state of harmony, but rather a process of change" in which societal action negotiates between changing human needs and environmental capacities (9, 44, 65). When it describes behaviors suitable for this dynamic vision of sustainable development such as limiting consumption and striving for increased efficiency, the Brundtland Commission encourages activities that fit under the principle of careful use (55–57). Finally, throughout the entire Report, the Commission tries to take a long-range view of actions and their consequences, a principle I call farsightedness and develop further in Chapter 4.

The Report's vision of sustainable development was seen as a compromise between the historical prioritization of growth and economic development and the relatively newly powerful environmental movement (Du Pisani 2006: 94). This compromise helped spark a massive movement which hoped to simultaneously achieve social equity, economic growth, and environmental quality. Indeed, after the publication of the Report, the popularity of the terms "sustainability" and "sustainable development" grew exponentially as scholars, policy-makers, and average citizens debated their meanings and merit and tried to implement these ideas in their actions.

Many critics of the Brundtland Report have focused on its ethical assumptions and implications. In particular, scholars and activists have often disagreed over whether the Brundtland Report's ethical foundations are wide enough or have been applied to sufficiently diverse moral communities. Critics, including those in *Beyond Brundtland*, claim the Report problematically focuses on poverty and population explosions as the cause for environmental destruction without giving adequate weight to and assigning adequate blame and responsibility to the wealthy who are tremendous polluters (Court 1990: 13–15). Others critique the Report for its concentration on Western standards of living and the instrumental value of the environment. They question whether equity can be properly defined only in reference to certain human lifestyles or even to humans in general (Court 1990: 15, Norton 1992: 99). While the Report occasionally suggests that moral obligations to nonhumans exist (World Commission on Environment and Development 1987: 57), the opposition counters that such references are tacked onto the Report's claims rather than penetrating each proposal. Thus, the World Council of Churches modified its language about its ethical vision from "Justice Participation and a Sustainable Society" to "Justice, Peace, and the Integrity of Creation" in the late 1980s to avoid solely considering the environment in anthropocentric terms as the sustainability movement sometimes does (Wogaman 1993: 266, World Commission on Environment and Development 1987: 57, World Council of Churches, World Council of Churches. Central Committee 1975, World Council of Churches. Central Committee 1983: 9, World Council of Churches. Central Committee 1990). These critiques arise out of questions such as who gets to decide which ethical priorities shape sustainability discussions and how people with conflicting ethical visions can come to consensus over sustainable development. Critics also question whether the Brundtland Report's focus on economic growth and technological advancement can sufficiently address ecological problems, or whether a change in moral and cultural priorities is needed. Thus, they wonder whether its balance of technical and normative aspects of sustainability is appropriate and how to better relate these components. All of these critiques of the Brundtland Report highlight the fact that technical and normative aspects of sustainable development are intertwined as people question whether the sustainability movement prioritizes what they value.

Though these normative questions were at the heart of many direct criticisms of the Report, most responses to the Brundtland Report in the rapidly expanding sustainability movement focused and were framed as technical issues. For example, though the Brundtland Report was adamant about using the term "sustainable development," instead of "sustainability" to bridge the gap between development studies and the environmental movement, many expressed doubts about this hybrid term either because they were suspicious of development as too focused on the economic in and of itself, because it often aids "donors" more than "recipients" in developing countries, or because they questioned whether "sustainable development" was in fact oxymoronic. Others debated whether resources could be substituted for each other and still have sustainability (the debate between weak and strong sustainability) or the best definition of the relationship between economics and sustainability (Ayres 1997, Ayres and Nair 1984, Clark 1997, Common 1997, Daly 1997a, Daly 1997b, Opschoor 1997, Pearce 1997, Peet 1997, Simon 1996, Solow 1997, Stern 1997, Stiglitz 1997, Tisdell 1997). Indeed, clarifying these conceptual problems seemed a necessary precursor to articulating and implementing sustainable development policy and moving toward sustainable development for many in the late 1980s and 1990s (Bartelmus 1997: 198, Bartelmus 1999: 4–6, Cobb and Halstead 1994: 2–6, Daly et al. 1989: 62–84, England 2001: 218–19, Hueting 1991: 194–96, Hueting et al. 1992: 7–9, Lefever 1979: 45, Redclift 1987: 56–77, Taylor 1995: 12, van Dieren 1995, World Commission on Environment and Development 1987: 290).

2.2.2 For the Common Good

Herman E. Daly and John B. Cobb Jr.'s *For the Common Good* significantly contributed to the burgeoning sustainability movement as they worked to incorporate the insights of many disciplines including philosophy and ethics into their discussion of economics, sustainable development, and the common good. In this influential text, first published in 1989, Daly and Cobb clarified conceptual debates about the meaning and value of "development" associated with sustainability by distinguishing between development and growth. They also catalyzed the movement to monitor progress toward sustainable development by critiquing existing economic measures and ultimately, constructing a new measure, the Index of Sustainable Economic Welfare (ISEW) that included more environmental and social considerations. Examining these three elements will indicate the major ways that Daly and Cobb worked to integrate the technical and normative aspects of sustainable development and will enable an assessment of their influence on the movement.

A major debate about associating development with sustainability after the Brundtland Report hinged on the relationship of physical resources and development. Briefly, some scholars thought that economic development was the answer to all or most human problems while others recognized that economic progress was dependent on and limited by physical resources and a wide variety of values. This second group maintained that economic growth is only desirable as a means to improve quality of life, not as an end in and of itself.

Daly and Cobb stabilize this debate by recognizing that "sustainable growth" and "sustainable development" are erroneously used as synonyms, a trend that leads to logically impossible conclusions regarding economics and natural resources such as the possibility of an infinite supply of resources and their infinite substitutability. They solve this problem by carefully distinguishing between sustainable growth and sustainable development:

> "Growth" should refer to quantitative expansion in the scale of the physical dimensions of the economic system, while "development" should refer to the qualitative change of a physically nongrowing economic system in dynamic equilibrium with the environment. By this definition, the earth is not growing, but it is developing. Any physical subsystem of a finite and nongrowing earth must itself also eventually become nongrowing. Therefore, growth will become unsustainable eventually and the term "sustainable growth" would then be self-contradictory. But, sustainable development does not become self-contradictory.
> (Daly et al. 1989: 71–72)

Thus, Daly and Cobb suggest that sustainable developmental entails improvements in the economy that do not necessitate increased interaction with the environment. Rather, they claim that sustainable development can include all human activities and indicate qualitative improvements in quality of life without increasing human impact on the environment. In this way, followers of Daly and Cobb's terminology are able to avoid charges of physical impossibility while discussing sustainable development.

Throughout this analysis we see that Daly and Cobb intertwine normative and technical assessments of sustainability: they recognize the planetary limits to development (technical) and desire sustainable development which involves an improvement in quality of life (normative) without increasing human impact on the environment (requiring technical assessment, but guided by a normative vision).

Daly and Cobb's desire for the improvement and maintenance of the common good drove them to technically and normatively critique existing economic measures such as Gross National Product (GNP) that are used to assess a country's well-being and suggest an alternative, the Index of Sustainable Economic Welfare (ISEW). GNP, an economic index of the monetary value of goods and services purchased by households (Daly et al. 1989: 65), is often used as an overall measure of economic success and national well-being, an application beyond its original intended purpose. Yet as Daly and Cobb, among others, point out, GNP does not adequately count income (Daly et al. 1989: 70–71) or acknowledge the value of nonmarketed domestic work, leisure, and environmental services such as water filtration by wetlands (Daly et al. 1989: 454–59), all factors which significantly impact individual and communal well-being and sustainability. Their ISEW accounts for such issues, at least on a basic level, by explicitly identifying indicators according to their normative priorities. For example, they follow the WCC in thinking that the world needs a new vision of well-being that is "just, participatory, and sustainable" (Daly et al. 1989: 20). Their focus on justice leads them to prioritize income distribution so that they directly incorporate it into their index emphasizing "the plight of the poorest members of society" (Daly et al. 1989: 445, 465). In this way, Daly and Cobb advocate for and model the influence of normative concerns during the process of index development.

They argue that such explicit work is necessary because of the disciplinary development of economics, largely patterned after the quantitative laws of the sciences that led to a vision of economic life in which humans were individualistic rational actors rather than parts of complex communities with complex beliefs and values and the contributions of the natural environment to human well-being aside from land ownership was ignored completely (Daly et al. 1989: 85–120). This vision led to economic analyses largely disconnected from many features of human life and well-being in communities. Such abstractions, they argue, are too often taken as reality, limiting the understanding of and response to real world problems including environmental issues. To overcome the limits of this approach, Daly and Cobb advocate for a nondisciplinary economics focused on community well-being, opening economic analyses up to the influence of other disciplines including agriculture environmental studies, history, sociology, and philosophy to broaden the technical assessments included in economic theories and indexes (Daly et al. 1989: 123). They illustrate this approach in the development of the ISEW.

While Daly and Cobb make a strong case for a nondisciplinary economics and study of sustainable development, they leave a number of issues relatively unexplored. First of all, such work requires participants that are broadly trained and/or are willing and able to work with people of multiple disciplines as they did (Daly trained as an economist and Cobb as a theologian and philosopher). Yet disciplinary training programs and incentives (e.g. publication in disciplinary journals, tenure) often dissuade such endeavors.

Even if someone is interested in such work it is quite difficult to gain the necessary expertise. Thus, attempts to follow or extend Daly and Cobb's work to more thoroughly integrate norms into indexes face a number of systematic challenges.

Additionally, the fact that Daly and Cobb not only advocate for including norms and metaphysics in sustainability conversations but advocate for *a particular* view raises theoretical and practical problems. They maintain that a religious worldview is necessary for a move toward a sustainable community *and* support a particular view, namely a "theocentric undergirding of the biocentric perspective," rooted in Christian process theology (Daly et al. 1989: 401). They find this approach advantageous because it encourages people to recognize the limits of humanity and therefore of human models, a step toward avoiding the fallacies of much economic thinking in which models dissociated from community and ecosystems are taken as reality, abstractions that move analysis and action away from sustainability (Daly et al. 1989: 401–4). While these features of their perspective may be helpful for some, its potential utility will probably not convince people to adopt this worldview. Daly and Cobb admit that many different worldviews exist but do not explore in detail how their perspective will interrelate with others or how people of different worldviews can collaboratively develop sustainable well-being indexes to replace traditional economic measures. Thus, their work needs revision to appeal to cross-cultural audiences who are interested in sustainability indexes.

Admittedly, much of Daily and Cobb's work on religion in *For the Common Good* aimed at showing that Christianity does not have to lead to a disrespect for nature as many assumed at the time (Daly et al. 1989: 382–83). They needed some degree of specificity to do this. But now, twenty some years later, when it is much more widely acknowledged that religious and philosophical worldviews can contribute to environmental thinking (partially due to their work), we need to ask how the multiple perspectives existing in our world, including those not explicitly associated with a religion, can collectively influence visions of sustainability and the content of sustainability indexes. If we wait for everyone to share the same religious worldview as a foundation for sustainability ethics, sustainability efforts will be on hold for a long time, if not forever.

Unfortunately, possibly due to these unaddressed aspects of their interdisciplinary approach, sustainability studies did not build upon Daly and Cobb's vision of integration, though their definition of sustainable development and focus on indexes did significantly impact the movement as we shall see in the following sections.

2.2.3 Rio Earth Summit

Daly and Cobb's *For the Common Good* was not, however, the only influential document that considered an integrative approach to sustainable development.

The United Nations Conference on Environment and Development (UNCED) at the 1992 Earth Summit in Rio de Janeiro acknowledged that both technical and normative aspects of sustainable development were important to achieving it when it developed and adopted Agenda for the Twenty-First Century (Agenda 21) and the Rio Declaration on Environment and Development. These documents were intended to complement each other by focusing on technical and normative aspects of sustainable development respectively in order to implement the Brundtland Report's vision (Robinson et al. 1993: xii, 1). Yet a number of historical circumstances surrounding the response to these documents would lead to the supersession of the Rio Declaration by the Earth Charter, the focus on Agenda 21 as a technical document, and the increasing bifurcation of technical and ethical studies of sustainability. This division ignores the normative elements of Agenda 21, overlooks the necessity of integrating technical and ethical aspects of sustainability, and leaves a host of questions unanswered and unanswerable, inhibiting movement toward sustainable development.

At the Earth Summit, Agenda 21 was adopted by representatives from over 170 nations (Robinson et al. 1993: xiii, xxi). At the time it was "the world's most important soft law document," a class of documents that recommend but do not require governmental action (Robinson et al. 1993: xv, xxi). Agenda 21 bases its discussion of sustainable development on the definition from *Our Common Future*: "meet[ing] the needs of the present without compromising the ability of future generations to meet their own needs" (World Commission on Environment and Development 1987: 8), though the entirety of its more than 800 pages, and the debate that it sparked, can be considered a definition of sustainable development as it obliquely articulates its vision of sustainable development in order to construct a scientific, economic, and legal blueprint for how to measure it (Robinson et al. 1993: iii).

As a blueprint for action, Agenda 21 examines the social and economic dimensions of sustainable development in order to outline the role of governments and laws in reaching sustainable development. It explores connections between sustainable development and the relation of poverty and consumption; resource management, particularly pertaining to the atmosphere, land, oceans, and wastes; and the actions of major segments of the population including women, youth, indigenous people, and nongovernmental organizations. It itself does not produce new scientific or technical assessments; rather, it encourages and mandates such studies to further its goals. Insofar as Agenda 21 outlines and encourages financial, scientific, legal, and educational methods of studying sustainable development and implementing sustainable development goals, it encompasses technical aspects of sustainability.

The Earth Summit organizers did not intend Agenda 21 to stand alone as a technical blueprint for achieving sustainability as they followed the Brundtland Report's call for a new document outlining a universal normative vision on environmental issues by composing a document of norms, including rights and responsibilities, pertinent to sustainable development and environmental issues, a significant addition to international agreements about rights (Rolston III 1994: 267). States and to a lesser degree individuals are identified as the primary agents to enact such norms, though the Declaration notes that often marginalized groups including youth, indigenous people, and women can significantly contribute to sustainability initiatives. The Declaration primarily aims to "respect the interests of all and protect the integrity of the global environment and development system" through twenty-seven ethical principles (United Nations Conference on Environment and Development 1992). The Declaration prioritizes humanity from its first paragraphs where it states that "Human beings are at the centre of concerns for sustainable development." Within this anthropocentrism it emphasizes the needs of the most vulnerable humans, hoping to ensure that all have the necessary information and the ability to participate in the decision-making process, and that "the environment and natural resources of people under oppression, domination and occupation shall be protected" (United Nations Conference on Environment and Development 1992). The Declaration also prioritizes "conserve[ing] protect[ing] and restor[ing] the heath and integrity of the ecosystem," and "reduc[ing] consumption and enact[ing] solid demographic policies." These goals are, however, balanced by the Declaration's support for states' rights "to exploit their own resources" under international law and the UN Charter (United Nations Conference on Environment and Development 1992). Thus, the Rio Declaration highlights the responsibility of governments and individuals to promote the rights of individuals, particular groups of people, and nations with respect to their environment.

Yet many environmentalists and ethicists were highly critical of the Declaration, particularly because of its focus on the instrumental value of nature to humans, supposedly not leaving an option for valuing nature for its own sake; because of its inadequate attention to disparities in access to environmental goods and services among demographic groups; and because it emphasized technological fixes and western capitalism over a variety of solutions to environmental challenges (Brown 1994, Brown and Lemons 1995). Thus, in 1994, Maurice Strong and Mikhail Gorbachev led a grassroots movement to support a revised version of the Rio Declaration, the Earth Charter. Since its final form was reached in 2000 the Charter has been endorsed by thousands of groups and individuals worldwide (Earth Charter Initiative).

The Charter extends or supplements the Declaration in several key areas. Whereas the Declaration's language prioritizes the legalistic recognition of rights and duties in the quest for sustainability, the Charter also stresses the importance of developing new vision of life and global community. Similarly, the Charter explicitly distances itself from the ideals of capitalism and materialism as it maintains that "when basic needs have been met, human development is primarily about being more, not having more" (The Earth

Charter Commission 2000). It recognizes that spiritual and cultural traditions, environmental resources, health, peace, and relations with including but not limited to humanity, are all critical to well-being (The Earth Charter Commission 2000). As it diversifies its focus beyond physical and economic needs, the Charter also takes care to ensure that various groups of people including the poor, women, children, and indigenous groups are not only seen as resources to teach others about sustainability but also valuable in and of themselves. Similarly, it asserts that "every form of life has value regardless of its worth to human beings," a departure from the anthropocentrism of the Declaration (The Earth Charter Commission 2000).

Many environmental ethicists were involved in the Earth Charter initiative, writing and commenting on its drafts, advocating for it at a grassroots level, chronicling the process for academic audiences, and/or advocating the need for a new world ethic in general (Aiken 2001, Earth Charter Initiative, Hargrove 1997, Tucker 2004, Weiming 2001). Certainly some ethicists critiqued the Earth Charter project as a whole because they were wary of the idea of a global ethic as they focused on the particularities of individual ethical systems in general, because they thought the Charter was biased toward the West, because it did not align with their normative or cultural visions, or because they did not think a global ethic was a possible or strategic move to aid the environment. Yet critics were significantly engaged with the Charter and thereby largely ignored the Earth Summit documents.

Yet Agenda 21 itself implicitly reflected a number of normative priorities and even hinted at ethical methods appropriate for a multicultural world though it use of language from law, economics, environmental studies, and developmental studies to make its claims. As ethicists overlooked its ethical content they loosened connections between technical and ethical aspects of sustainability and sustainable development. Examining these trends illustrates how the technical and ethical were linked in Agenda 21 in addition to the linkage of these aspects of sustainable development through the pairing of Agenda 21 and the Rio Declaration.

The few places in which Agenda 21 overtly references ethical considerations shed light on its commitment to a diversity of values. Most common, mentioned at least ten times throughout the document, is its recognition that signatories and their constituents may hold diverse perspectives on reproduction (Robinson et al. 1993: 28, 42, 45, 47, 51–55, 57, 68, 72, 75, 494–95). Divergent values are also acknowledged as Agenda 21 maintains that the prevention and treatment of diseases should resonate with the social, religious, and ethical values of individuals and communities (Robinson et al. 1993: 57, 61–62). In general, Agenda 21 supports any norm that could promote sustainability. For instance, it maintains that states should "reinforce both values which encourage sustainable production and consumption patterns and policies which encourage the transfer of environmentally sound technologies to developing countries" without naming particular values (Robinson et al. 1993: 37, 39). Quite rarely, Agenda 21 identifies specific ethical practices consonant with sustainability in particular cultures, such as ḥimā, traditional Islamic land sanctuaries. (Robinson et al. 1993: 156). These references demonstrate that Agenda 21 supports diverse ethical values at least where such values promote actions that align with its goals. Given the legal and technical focus of Agenda 21, it is not surprising that it would limit its discussion of ethics to focus on the roles of government, laws, and technology to move toward sustainable development rather than on ethical assessments. Yet, its very carefulness calls for additional study because Agenda 21 does not delineate how the diverse ethical traditions can work together. I see this as a point when policy experts envisioned an adaptable type of ethics that was neither absolutist nor entirely relativistic but did not have the terminology, expertise, or space to develop such an ethical method. Thus, I view it as a small step toward an ethical method, a place where professional ethicists have a unique role to play in the sustainability debate and the creation and implementation of sustainability indicators.[2]

Except for the few explicit references to ethics and values as mentioned above, moral commitments are implicit in Agenda 21. They can, however, be identified by analyzing its word choice and policy recommendations. Through this process, undertaken below, we see that acting sustainably according to Agenda 21 involves, at minimum, valuing a multifaceted ethical approach and following the ethical principles of farsightedness, adequate assessment of the situation, adaptability, cooperation, efficiency, responsibility, and equity. Though not all of these principles are traditionally considered ethical, they function in Agenda 21 as broad interrelated guidelines for ethical action that may take different forms according to environmental and social settings.

Farsightedness, concern for the long-term temporal and long-range spatial consequences of actions, was a central motivation for the development of Agenda 21. World leaders recognized that seemingly small actions can have severe, long-lasting environmental and social effects and recognized the ethical significance of these distant outcomes. The general commitment to farsightedness is seen throughout the document, as when it states that "Agenda 21 addresses the pressing problems of today and also aims at preparing the world for the challenges of the next century" (Robinson et al. 1993: 1). Agenda 21 also applies farsightedness to particular aspects of human activity such as agriculture, healthcare, and industry (Robinson et al. 1993: 233–34). In these ways, Agenda 21 encourages its signatories to consider distant and long-term consequences of human action and cultural and ecosystem changes.

Recognizing that the state of ecosystems and societies and their interaction influences how sustainable development can be conceived of and attained, Agenda 21 emphasizes adequate data assessment in its structure and treatment of specific issues. Each section utilizes current, thorough, and trustworthy information about the systems in question. The reliance on data continues throughout its articulation of objectives and methods of achieving these objectives as technological development, research into the state of the environment, and education are frequently mentioned means of reaching sustainability objectives (Robinson et al. 1993: 9–10, 29, 35, 37, 42–43, 46, 53, 58–59, 63–65, 70, 73–79, 89, 96, 99–100, 108, 119–20, 124–25, 133–34, 143–44, 158, 163–65, 169–71, 175–77, 179–82, 187–89, 192–93, 196–97, 200–1, 204–5, 208–9).

Agenda 21's attention to the adequate assessment of data is also evident in its treatment of specific issues. For example, Agenda 21 maintains that while poverty results in certain kinds of environmental stress, the major cause of the continued deterioration of the global environment is the unsustainable pattern of consumption and production, particularly in industrialized countries, which is a matter of grave concern, aggravating poverty and imbalances (Robinson et al. 1993: 32). Paying attention to the ways both consumption and production lead to environmental destruction is a more thorough assessment of the situation than *Our Common Future*'s emphasis poverty's contributions to environmental degradation.

Agenda 21's emphasis on adequate research and data is not surprising for a major international legal document concerning the environment and development. What may be surprising is the fact that I name such a priority an *ethical* principle. Ethicists who focus on policy analysis or applied ethics may think that this is unnecessary because paying attention to the situation is an inherent part of ethical practice. Yet, identifying adequate assessment of the situation as an ethical principle distinguishes Agenda 21's ethics from several strands of environmental literature. First, some ethicists focus upon abstract ideals to the point that they only give cursory attention to the geologic, environmental, biological, economic, political, or social details of a situation. In contrast, maintaining that the details of a situation are significant in ethical reflection and action is an ethical position. It is also an ethical position when compared to the common assertions that environmental issues are strictly moral problems that can be fully solved by changing societal values or alternatively that environmental issues are strictly technical problems resolvable through science and technology alone, without any consideration of values.

Yet the move to name adequate assessment of the situation an ethical principle is not merely a rhetorical move to distinguish Agenda 21's ethics from that of other parts of environmental literature. Adequately assessing the situation emerges as an ethical guideline when Agenda 21 is read as a possible source of ethics. Agenda 21 assumes that ethical priorities emerge out of the context in which people find themselves; they are not just abstract ideals. Thus, to develop and implement a blueprint for sustainability one must understand the technical details of the situation in which movement toward sustainability is desired and the pertinent values. Agenda 21 recognizes that technical capacities to sustain a system will not be implemented if they obviously yield results that are morally reprehensible to the implementers. Similarly, ethical ideals about what should be sustained, if distanced from technical realities, will never be implementable or achieved. Thus, to move toward sustainable systems, the goal of Agenda 21, ethicists must be attentive to the best available knowledge about systems they desire to sustain. To refuse to acknowledge, or seek out, the best knowledge about an environmental situation and its relationship to human society is to act unethically. As knowledge is shaped by presuppositions about what is valuable, a significant dialogue between technical disciplines and ethics is necessary to understand and value any situation. Thus, adequate assessment of the situation includes ethical content (that it is right to know the situation) *and* a method for enacting this principle (collaboration with relevant experts).

Another common ethical principle in Agenda 21 is adaptability. Adaptability appears when Agenda 21 outlines how activities for sustainable development are to be managed. It emphasizes that governments should work at the appropriate level to address questions of sustainable development "in collaboration with the national and international scientific community and with support of appropriate national and international organizations" (Robinson et al. 1993: iii). Agenda 21 also prioritizes adaptability as it promotes shaping environmental laws to the financial, educational, and research capacities and infrastructure of a particular country; the uneven resources of political entities; and the need to respond to environmental and social change (Robinson et al. 1993: xxv, xxvii–lxxx, lxxxi, lxxxiv, lxxxvii, lxxxix, 3–4). For instance it recognizes that different types of policies, short-term goals, and actions are needed for developing and developed countries and for different constituencies within countries (Robinson et al. 1993: 2, 6, 20, 24, 28, 34, 115–16, 351). It also notes that the policies it sets out must be adaptable to various "personally held values," and "ethical and cultural considerations" (Robinson et al. 1993: 52). A typical example of the language used to describe the need for this type of adaptability includes the injunction to:

> Implement, as a matter of urgency, in accordance with country specific conditions and legal systems, measures to ensure that women and men have the same right to decide freely and responsibly on the number and spacing of their children and to have access to the information, education and means, as appropriate, to enable them to exercise this right in keeping with their freedom, dignity and personally held values, taking into account ethical and cultural considerations.
> (Robinson et al. 1993: 28)

With these carefully worded phrases Agenda 21's authors tried to take account of the variety of cultural and ethical beliefs about reproduction while advocating action in the face of the implications of rapidly increasing global populations.

Adaptability is an ethical principle as it arises from the recognition that universal policies are often harmful to the environment and/or groups of humans. Adaptability is also an ethical principle because it guards against the hubris of believing that one knows enough about environmental situations to make universal policy decisions for all places and times. Adaptability also indicates something about the ethical method assumed in Agenda 21: broad ethical ideas are endorsed with the understanding that they can be implemented by adherents of widely diverse cultural and ethical traditions. Additionally, since adaptability is to be applied to environmental law, and Agenda 21 is a soft law document, then it suggests that the normative and technical recommendations of Agenda 21 should be open to revision as new situations, perspectives, and knowledge emerge. Thus, like adequate assessment of the situation, adaptability includes both ethical content (to act and make decisions in ways that can be applicable to many contexts) and ethical method (to recognize that ethical content should be revisable over time).

As seen in its discussions of adaptability, Agenda 21 mandates that humans have a duty to cooperate with each other in their sustainable development efforts. Indeed, its authors indicate that cooperation is the only feasible and acceptable means of achieving sustainability given the interdependence of all people, economies, societies, and ecosystems in today's world (Robinson et al. 1993: xxxvii–xxxviii, 4–23). After all, the dictatorial imposition of measures to achieve sustainable development, an extreme alternative to cooperation, would counter Agenda 21's emphasis on the rights of individuals to participate in decision-making processes according to their cultural, religious, ethical, and personal values (Robinson et al. 1993: 2, 26, 28, 42, 45, 47, 51–55, 57, 68, 72, 75, 492–95, 620–26). Three major types of cooperation are emphasized in Agenda 21: developed countries sharing technical information with developing countries (Robinson et al. 1993: 7–10, 29, 37, 96, 143, 620); coordinating governmental and nongovernmental resources (Robinson et al. 1993: 72, 137–38, 140, 157, 162, 230, 267, 387, 392, 471, 484, 567); and building consistent international systems of environmental law (Robinson et al. 1993: 620–26). To a lesser degree, cooperation is also supported in Agenda 21 as it advocates employing traditional and newly developed methods of land use and biological resources.

The efficient allocation and use of natural resources and information is also encouraged by Agenda 21 with respect to land, water, energy, timber, biodiversity, education, healthcare, shelter, and the reduction of wastes (Robinson et al. 1993: 4, 33, 36–37, 139–40, 152, 166, 173, 280). Agenda 21 claims that humans "need to change consumption and production to reduce environmental stress and meet basic needs" and need to change values to ensure that we move toward sustainable development (Robinson et al. 1993: 34, 37, 166). Thus, Agenda 21 promotes both efficiency and decreased consumption. However, when it advocates particular types of actions, Agenda 21 emphasizes technological fixes over ideological changes or consumption decreases within the current technological regime (Robinson et al. 1993: 139–40). (A notable exception is its concern to reduce overpopulation.) Agenda 21 may focus on technology because its authors wanted to follow the division in Western thought between law and ethics, because ethics was considered too controversial to discuss in depth, or because its writers did not have enough ethical expertise. Whatever the reason, Agenda 21's stated commitment to total frugality yet focus on efficiency requires further ethical reflection.

To enable its other priorities when moving toward sustainable development, Agenda 21 emphasizes responsibility. Based on the interdependence of all people, economies, societies, and ecosystems in today's world and the commonsense idea that actions have consequences, Agenda 21 promotes at least two types of responsibility: admitting that we humans have contributed to environmental destruction; and recognizing that we need to change our actions to slow or reverse environmental destruction (Robinson et al. 1993: 32, 140, 142, 152, 161, 184, 212, 253, 263, 265, 309). Agenda 21 notes, for example, that people have caused air and water pollution, deforestation, species extinction, and desertification (Robinson et al. 1993: 32, 140, 142, 152, 161, 184, 212, 253, 263, 265, 309). It maintains that using new technology and overconsumption have led to a majority of environmental problems, though poverty has also played a role in damaging the environment (Robinson et al. 1993: 32, 184). Thus, Agenda 21 supports the first type of responsibility.

Following the second aspect of responsibility, Agenda 21 recommends modifying behaviors, values, and standards of living to achieve sustainable development (Robinson et al. 1993: 29, 35, 53, 108). Most of Agenda 21's in-depth examples of how these changes should come about rely on bringing the technology and economic systems of developed countries to developing countries. Unfortunately, it does not reconcile the tension that exists between promoting such development and the fact that these types of development helped exacerbate the environmental, social, and economic situation that we are in today (Pereira 1997: 74–79). Despite this critique, Agenda 21 does include injunctions to take responsibility for our past and future actions, the second type of responsibility.

While the ethical principles discussed above certainly pervade Agenda 21, its treatment of equality has received the most attention among ethicists. The vision of equality for all people is emphasized through Agenda 21's devotion of an entire section of chapters, "Strengthening the Role of Major Groups," to the roles of children, youth, indigenous people, nongovernmental organizations, local authorities, workers and unions, business and industry, the scientific and technical communities, and farmers in obtaining sustainable development (Robinson et al. 1993: 492–545). Equality for these people is not a goal that is merely added onto the rest of the document. Rather, equitable access to physical resources and the political process is emphasized in discussions of trade, global cooperation and economics, poverty and hunger, consumption, natural resources, the diverse needs of developing and developed countries, and access to the power of governmental and nongovernmental organizations (Robinson et al. 1993: 2, 5–8, 20, 24, 26, 28, 32, 34, 45, 50–54, 58, 60, 62, 66–67, 73, 83, 89–90, 115–16, 229, 351, 492).

While Agenda 21 gives special attention to groups of women, children, the poor, developing countries, and indigenous people, its language suggests that all groups should be treated equally (Robinson et al. 1993: 26, 46, 50). The ability to extend equality from Agenda 21's specific examples to any group in need of special attention is significant since its analysis of the proper treatment of "major groups" ignores one that has disproportionally borne the burdens of environmental destruction: racial and ethnic minorities (Robinson et al. 1993: 23, 492). Given the lacuna in Agenda 21 with respect to this issue, many ethicists at a 1994 conference on ethics in Agenda 21 (the major ethical commentary on the document) argued that more attention to issues of justice, for example who benefits or is harmed by environmental degradation and who participates in decision-making about it, is needed in general and will be critical to implementing Agenda 21 (Bullard 1994a: 59–61, Heredia 1994: 123–27, Heyd 1994: 131, Ott 1994: 219, Paden 1994: 261–63, Rolston III 1994: 270–80, Warren 1994: 321, Weiss 1994: 362). Except for Heyd, these scholars, however, do not claim that Agenda 21 itself is unethical with respect to justice unless they argue that it is unjust because it does not recognize the intrinsic value of nonhuman biota (Heyd 1994: 133, Katz 1994: 153–55, Paden 1994: 235–37, Sagoff 1994: 289–97, Tucker 1994: 316). Ethicists focused so much on the anthropocentric tendencies of Agenda 21 that they did not, except for Rolston, acknowledge the ways that it advanced a new international vision of justice by catalyzing and articulating two new, interrelated, justice principles: "an equitable international economic order" and "protection of the environment" (quoted from Jones, in Rolston III 1994). Thus, the vast majority of ethicists focusing on justice, biocentrism, or intrinsic value quickly moved past Agenda 21 to ethical conversations, neglecting the technical and ethical elements that Agenda 21 and the Rio Declaration represent and Agenda 21's focus on monitoring sustainable development progress. Unfortunately, this move overlooks the contribution of the Summit to the international ethics discussion, particularly the way it weaves concerns for equity (and other ethical concerns) among humans throughout its technical analysis.

Throughout this examination of Agenda 21 we see that its technical understanding of a dynamic, complex, multicultural world in which groups of people and countries have different environmental histories, resources, and cultures is interlinked with normative principles including adequately assessing the situation, adaptability according to time and situation, farsightedness, equality, responsibility, cooperation, and efficiency as well as an ethical method open to specification by a diversity of traditions. Yet ethicists, after a brief examination of Agenda 21 in 1994, tended to ignore it in favor of the Earth Charter, contributing to the bifurcation of the sustainability movement, especially regarding indicators.

2.3 Diversification after Rio

Agenda 21 sparked a massive increase in sustainability studies and policy-making as people focused on defining sustainability and sustainable development for particular sectors of society such as agriculture or energy and developing methods of monitoring progress toward sustainability for countries, cities, businesses, or specific industries. All of these initiatives focus on the technical and are hindered by their lack of explicit attention to ethics. In contrast, another group of scholars focus on philosophical and ethical aspects of sustainability but as they emphasize the abstract realm of ethical theory they do not sufficiently examine how sustainability norms can be implemented in indexes. Thus, they often neglect questions critical to the implementation of sustainability initiatives. The aspects of the sustainability movement which do a better job of recognizing the connections between ethical and technical aspects of sustainability still do not adequately relate these connections to indexes. Examining these segments of the post-Rio sustainability movement will demonstrate the degree to which sustainability studies has been bifurcated and identify the questions left unanswered and unanswerable by such methods.

2.3.1 Sectorial definitions

After the Brundtland Report's general vision of sustainable development and Agenda 21's blueprint for achieving it, scholars and activists recognized that they needed more specific definitions of sustainable development and sustainability to guide their study and activism, particularly for specific segments of society (e.g. agriculture or energy use), a particular group, and/or a particular spatial and temporal scale. Two representative definitions, the United Nation's Food and Agriculture Organization (FAO)'s definition for agricultural sustainability and Jefferson Tester et al.'s definition of sustainable energy, demonstrate that 1) such specifications of sustainability tend to prioritize technical elements though ethical elements also pervade the definitions and 2) this emphasis hinders the application of the definitions.

In 1998, the United Nation's Food and Agriculture Organization (FAO) devised a definition of sustainable development pertinent to its activities:

> Sustainable development is the management and conservation of the natural resource base and the orientation of technological and institutional change in such a manner as to ensure the attainment and continued satisfaction of human needs for present and future generations. Such sustainable development (in the agriculture, forestry, and fisheries sectors) conserves land, water, plant and animal genetic resources, is environmentally non-degrading, technically appropriate, economically viable and socially acceptable.
>
> (FAO Council 1988, Tschirley 1997: 221)

Though short, compared to the full Brundtland Report or Agenda 21, this definition reveals a significant amount about the ideals it is founded upon. Social and economic viability, as well as components of justice are highlighted. It is presumed that genetic resources are necessary to avoid catastrophe from agricultural pests given the predominance of monocultural farming in contemporary agribusiness; a collapsed economy will not allow for the safe fulfillment of people's needs. Additionally, the FAO definition assumes the principle of responsibility because without responsibility, there would be no impetus to care or take action.

The ethical vagueness of the FAO definition is not that surprising given the technical aim of the FAO. Yet it hinders the discernment of 1) the scope of the FAO's ethical claims and 2) how they are to be applied. For example, while the FAO definition appears to be anthropocentric, it is not at all clear if and how it supports equality among people within a generation or how it would apply its terms such as "social vitality" or "technical appropriateness." Furthermore, the FAO definition and the discussion surrounding it do not explain how ethical principles are to be prioritized if they conflict or how its normative priorities draw upon or relate to the diverse technical and cultural traditions around the world. Of course, the FAO documents are intended to be technical documents, not a fully articulated ethical system, yet as we can see, even "technical" documents about sustainability raise considerable questions about ethical content and methods that require reflection to fully implement the technical concerns.

This trend persists in definitions of sustainable development articulated with respect to particular sectors of life for educational purposes, as is illustrated in the definition of sustainable energy by Jefferson Tester in *Sustainable Energy: Choosing among Energy Options*. Though Tester's book is a technical introduction to 1) methods of extracting, refining, and using various fuels to yield useable energy and 2) the technical and environmental limitations of each source, it has a very normative definition of sustainable energy: "a dynamic harmony between the equitable availability of energy-intensive goods and services to all people and the preservation of the earth for future generations" (Tester 2005: 8). Tester's definition, like so many in discussions of sustainability, prioritizes actions that enable equity among people today and some sort of balance between these considerations and those of future generations.

Yet Tester and his colleagues incorporate normative ideals of harmony and equity directly into the definition without describing either term, discussing why they these ideas are necessary qualifications of sustainability to meet the needs of the future, exploring whether they are sufficient normative components of sustainability, or explaining how one should prioritize these elements with respect to other aspects of the definition. Such limitations make it difficult, if not impossible, to know when progress toward equity or harmony, and thus sustainable energy, according to Tester, is made. Prioritizing harmony makes the application of this definition particularly difficult if harmony is not specifically defined or at least generally discussed. It can all too easily be understood as indicating an environment in which nothing bad, unfortunate, or disastrous has a place, yet this is an inaccurate description of the world. The living die. Natural disasters upend societies. Given these facts, attempts to use harmony must be carefully constrained to prevent the idealization of environmental states. Yet there is a significant disconnect between the ideal vision of sustainable energy advocated by Tester and the educational tasks set forth in the text as a whole. Students are taught about coal mining, how nuclear plants work, and the wind energy potential of various places in the US, among other topics. They are, however, not given any methods for evaluating energy sources against each other despite the premise of the book, as indicated in the subtitle, "Choosing among Energy Options." Making normative decisions about the best energy sources cannot be made using "pure facts" alone.

Clearly then, focusing on the technical without attending to the ethical does not by itself enable people to put theories of sustainability into practice in policy-making or educational endeavors. To do so, technical and ethical considerations need to be explicitly acknowledged as mutually influential and accompanied by methods of articulating and partnering sustainability ethics with the relatively well-developed technical studies of sustainability from economics and the sciences. Unfortunately, scholars developing methods of monitoring progress toward sustainability have not yet done such combinatory work.

2.3.2 Monitoring and reporting

After Agenda 21, governments, nongovernmental organizations, businesses, and academics not only wanted to articulate sectorial definitions of sustainable development but also aimed to develop and implement methods of monitoring progress toward sustainability in order to help articulate sustainability goals and shape and evaluate sustainability policy. Building on the work of economists including Daly as well as the work of ecological scientists, demographers, and other social scientists, hundreds of such indexes were developed in the 1990s and early twenty-first century. Yet while these indexes were certainly influenced by normative visions of sustainability, such concerns were rarely explicitly or significantly incorporated into methods of monitoring and reporting sustainability. A full ethical analysis of particular sustainability indexes will wait for Chapter 5, here an exploration of the general trends within this monitoring movement, particularly the relative lack of attention to ethics, will reveal the limitations of such approaches.

After Agenda 21, guidelines and frameworks for the development of sustainability indexes became a significant research area. Index developers soon focused on a three-dimensional approach (the environmental, economic, and social), to encompass critical aspects of sustainability and draw upon the expertise of the environmental scientists, economists, and other social scientists that contribute to sustainability studies (Obst 2000). Theorists also classified types of indicators as in the popular Driving-Force–State–Response framework (and its variants) to conceptualize the complex, dynamic issue of sustainability (Forstner 1997, Mortensen 1997). Additionally, theorists discussed the technical challenges to identifying indicators and compiling them into indexes such that index developers would be aware of the trade-offs of multiple approaches (Bell 1999, Bell and Morse 2003, Gibson 2005, Hák et al. 2007, Moldan et al. 1997). While these initiatives were foundationally driven by normative concerns (the desire for a sustainable environment, economy, and social life) and sometimes recognized the potential contribution of ethicists to the subject, generally index theory has centered on technical, quantitative issues, a focus continued in the development of individual indexes.

Environmental and ecological studies have also contributed significantly to such indexes. For instance, technical studies of what could be sustained draw on ecological methods including methods of monitoring pollution rates; levels of accumulated pollution in air, soil, or water and the degradation of natural resources such as soil nutrients or forests; as well as environmental impact assessments. Ecological concepts such as carrying capacity have also influenced methods of monitoring sustainability. Carrying capacity refers to the ability of an ecosystem to provide for a population (e.g. with food, water, shelter) or assimilate waste. If the carrying capacity of an ecosystem is exceeded, populations will decline, possibly precipitously. Mathis Wackernagel, with assistance from his doctoral advisor, William Rees, built upon the idea of carrying capacity to develop the notion of Ecological Footprint, an index of the amount of land required to support a human given his or her lifestyle. If one's ecological footprint is high enough that the Earth could not support the entire human population if it had such an ecological footprint then one's lifestyle is deemed unsustainable (Rees and Wackernagel 1998, van den Begh and Verbruggen 1999). While the ecological footprint calculations explicitly focus on technical assessments of sustainability they also involve many normative assumptions. For example, Wackernagel and Rees presume that it is ethical to change one's lifestyle to ensure that one does not live in a way that is impossible for everyone to live while it would be unethical to kill off a large portion of the world's population to ensure that humans have a high quality of life within the world's carrying capacity. These claims could be based on a number of normative ideas including the categorical imperative or belief that all humans have rights or deserve care. These possible ethical foundations are not, however, discussed in their work.

Drawing upon environmental and economic methods of monitoring sustainability, hundreds of indexes for local, state, and national governments have been developed since Agenda 21. While some tend toward the academic exercise and have not been widely implemented, many others have been developed by or for particular governmental bodies. For example, Canada, the Netherlands, the United Kingdom, Costa Rica, China, Brazil, the Phillipines, and Austria have all developed and used sustainability indexes (Moldan et al. 1997: 299–378). While local governmental indexes may be more attentive to community values (e.g. clean parks) than national indexes, even they do not necessarily integrate ethics into their indexes in a significant way given the state of the literature and the predominance of data on economic and environmental factors.

Sustainability indexes have not, however, only been constructed for use at the governmental or individual level. In the late 1990s and into the new millennium businesses began developing and implementing their own sustainability indexes to align with the general public's increasing focus on sustainability. Certainly a few earlier pioneers of environmental responsibility within the business world, such as 3M, included some sort of monitoring effort in the mid-1970s to mid-1980s (Shrivastava 1995: 945). It was not until the 1990s, however, that there was a more widespread move to go beyond compliance and disaster preparation to focus on sustainability as a means to aid society and the environment while making savvy business decisions (Pane Haden et al. 2009: 1044). More businesses began to recognize that they could profit through environmental initiatives, typically through increasing the efficiency of their use of materials and energy and by letting consumers know of their efforts and accomplishments to create demand for their "environmentally responsible" products and services. Businesses at the turn of the century and beyond have increasingly studied ways in which their business and environmental goals could work hand in hand as they move beyond mere eco-efficiency in their products to linking the economic, environmental, and societal well-being of their company and society at large (Nattrass and Altomare 2001: 14–17, Pane Haden et al. 2009: 1045). To achieve these goals, many corporations started their own monitoring initiatives.

In 1994, R. H. Gray argued that the new wave of monitoring and reporting for sustainability should focus more on justice. Specifically, it should pay attention to the needs and desires of all people, not just company owners and stockholders, because sustainability involves time scales longer than any annual report and because "personal values are involved in delineating sustainability" (Gray 1994: 25, 30, 33). Thus, he argued that reporting methods should look at more than short-term economic factors and should be distributed more widely. For example, social sustainability indicators linked to business may monitor the number of jobs created and sustained over a period of time or whether and how the business collaborates with local communities to meet their social, cultural, and environmental needs.

While Gray did not see much evidence of this reporting in 1994, reporting for diverse stakeholders and community–company dialogue has increased significantly since then. Coupled with a more comprehensive vision of humanity in new indexes is an increase in the use of ecological monitoring methods based on ecosystem integrity and ecological footprinting into business assessments of sustainability. For example, as of 2006 nearly 80 percent of the Standard and Poor's 100 companies had social and environmental performance websites. The Global Reporting Initiative, a voluntary method of sustainability reporting, "has been implemented by more than 60 percent of the Global 1000 Companies, a plethora of nongovernmental organizations (NGOs) including the United Nations; and thousands of small to medium enterprises (SMEs)" (Gingerich 2010: 244, Hitchcock and Willard 2009: 10, 12). These companies and many others use techniques such as triple bottom line accounting which examines social and economic factors in addition to standard economic measures, or the Natural Step process in which businesses envision how they can simultaneously meet environmental and business goals and act to implement these visions (Nattrass and Altomare 2001). While these reporting initiatives are sometimes considered greenwashing, obscuring deeper environmental problems caused by businesses and capitalist economic structures with a veneer of eco-friendliness, at minimum they indicate the depth to which sustainability rhetoric has influenced business practices and consumer expectations.

Though methods of monitoring sustainability are most heavily influenced by and focused on technical methods of alluding to economic and environmental analysis, normative claims certainly pervade these accounts. Yet relatively little work has been done to determine how exactly to substantively incorporate diverse, possibly contradictory values into business indexes, as will be necessary to achieve sustainability and support for the businesses in this ethically diverse world. Another significant question for business sustainability indexes, or indeed, any indexes, is whether the values of the group using and affected by decisions contributed to the index.

Since the late 1990s indexes drawing on economic, environmental, and social analyses for use in environmental assessments, business, and governmental contexts have been developed at an exponential rate. While these indexes certainly have been developed at least in part in response to normative claims, as we have seen, they have focused primarily on economic and environmental analyses. Thus, considerable work will be needed to integrate normative and technical aspects of sustainability in sustainability indexes.

2.3.3 Ethical analyses

Ethical and philosophical investigations of sustainability often take technical assessments, such as the scientific consensus around climate change, as their starting point, quickly moving to abstract theoretical analyses and rarely, if ever, interacting with the theory and practice of developing and critiquing sustainability indicators. Instead, ethicists have focused on definitional issues of sustainability, the relationship of sustainability to other norms and to philosophical and religious traditions, and developing normative concepts such as intra and intergenerational justice that arise in sustainability discussions. The few exceptions to these trends discuss sustainability policy-making or monitoring sustainability in general terms, but do not interact with technical aspects of sustainability to the extent that they participate in the construction of or critique of sustainability indicators. Thus, ethical investigations of sustainability follow general trends of sustainability studies in which the integration of technical and normative aspects of sustainability, especially regarding sustainability indexes, is limited.

From early in the sustainability movement, many normative and philosophical studies of sustainability have focused on definitions of sustainability.

As previously discussed, the Brundtland definition and Report has received wide critiques from ethicists for focusing more on poverty than the responsibility of the wealthy, for being too anthropocentric, and for focusing too much on economic growth and technological fixes (Court 1990: 15, Norton 1992: 99, Wogaman 1993: 266). While many of these critiques were articulated in the years directly preceding the Brundtland Report, the continued prominence of this definition, and lack of satisfaction about alternatives, still sparks some discussions of the definition of sustainability nearly twenty-five years later (Callicott 2010: 60–61). Given such concerns, some ethicists have formulated their own definitions of sustainability (Engel and Engel 1990: 10, Gray 1994: 7–28, Norton 1992: 98, Shiva 1992: 188–92, Verburg and Wiegel 1997: 249). Responding to the proliferation of sustainability definitions and the many ways in which popular definitions such as the Brundtland definition are employed, philosophers have debated whether it is preferable to have one, multiple, or adaptive definitions of sustainability (Barrett and Grizzle 1999: 25, Norton 1999, Palmer 1992: 182, Redclift 1993: 4–11, Rydin 1999, Shiva 1992). While early discussions favored one definition, or at minimum, a unified definition for a sector of society, more recently, some scholars have noted that striving for complete definitional clarity may limit the sustainability movement's ability to take action on sustainability issues because it may never move past definitional issues or may obscure important normative elements in the process of consolidating definitions. Indeed, Lucas F. Johnston and Samuel Snyder, at least, argue that it is more important to identify the working definitions of sustainability that emerge in different institutions or communities as an illustration of their core beliefs than to develop an overarching definition of sustainability for, at minimum, this process will aid the understanding of and work toward sustainability in that community (Johnston 2013, Johnston and Snyder 2011).

Similarly, many ethicists study the relationship of sustainability to preexisting norms, virtues, or ethical principles and examine its resonance with preexisting philosophical and religious ethical systems (Barbour 2000, Hamed 2003, White 1994). While some see sustainability as a return to or development of ancient moral ideas, many see it as responding to significantly new environmental and social challenges. Some posit that sustainability is one of many important norms, virtues, or principles necessary for a contemporary ethic (Nash 1991: 64) while others consider sustainability to be the overarching ethical principle under which all other ethics fall (Pezzey 1992: 323, van Wensveen 2000: 132).

Predominant among the new normative and philosophical challenges of sustainability are the vast spatial and temporal scales involved in anthropocentric environmental impacts including climate change. Consequently, inter and intragenerational justice have captured the attention of many ethicists as they try to determine whether or not humans have a responsibility for or duty to people, and possibly biota, that are distant from them in space and time (Agyeman et al. 2003, Caney 2010a, Caney 2010b, Norton 2003b, Norton 2005: 332–55, Parfit 2010, Shue 2010a, Shue 2010b). Inspired by the importance of spatial and temporal scales, Norton has explored how to connect various scales in natural, economic, and policy-making with the scale at which particular values operate (Norton 2003a, Norton and Ulanowicz 2003).

Theoretical studies of the importance of spatial and temporal scales to sustainability ethics and philosophy, definitional studies of sustainability, and examinations of the relationship of sustainability to longstanding ethical systems typically begin with a recognition of technical assessments of sustainability. For example, authors often start their texts with statistics about the impact of growing human populations and industrial lifestyles on biodiversity, water quality, or climate. They then typically consider the ethicality of various types of actions given these initial conditions. Yet frequently, this engagement with technical assessments of sustainability primarily serves as a jumping off point for theoretical discussions. While these theoretical discussions are certainly needed given the many new challenges of sustainability ethics, focusing on theory is not a sufficient response.

Studies focused on the ethical dimensions of particular sustainability issues such as water, energy, or agricultural sustainability, which have proliferated since the 1990s, do, however, typically engage more deeply with technical studies. Some authors focus on the overall ethical issues including the power dynamics of these industries (e.g between industry and consumers, between large companies and small producers) or the relationship of water, food, or energy to quality of life. Yet, many, if not most, focus on case studies of particular places, communities, or companies to inform their analyses. Thus, in Kristen Schrader-Frechette's examination of nuclear waste, she explores the temporal scales and technologies necessary for containment, the health risks of exposure to nuclear waste for humans and biota, and the political struggles about nuclear waste disposal as well as theoretical analyses of inter and intragenerational justice (Shrader-Frechette 1991, Shrader-Frechette 2002). A number of other texts deeply examine ethical facets of energy use, but none explores the role of ethics in index formation or substantially addresses the challenges of cross-cultural ethics for sustainability (Barbour et al. 1982, Martin-Schramm and Stivers 2003). Similar trends are evident in ethical analyses focused on agricultural sustainability (Thompson 2010).

Though normative studies of particular sustainability issues engage more substantially with technical studies than other branches of sustainability ethics, studies focused on energy, water, or agricultural sustainability ethics do not engage with those technical aspects of sustainability related to monitoring progress toward sustainability. In other words, ethicists have typically developed theories of what is moral with respect to environmental, economic, and social dimensions of sustainability in general, but have not significantly worked with policy-makers or index developers to determine how these theoretical discussions could be integrated into concrete indexes. Thus, there is a significant gap between ethical reflections on sustainability and operationalizing these theories in policy-making and assessment. There are some exceptions to these trends, but all leave significant questions unaddressed.

Daly and Cobb's *For the Common Good* and other ethical critiques of economic indicators certainly set the stage for the contemporary ethical analysis of sustainability indexes. Yet because this work focuses primarily on economic questions and was undertaken in the late 1980s and early 1990s it does not examine the full scope of environmental and social issues related to sustainability or engage with the recent developments in sustainability activism and scholarship. Thus we must look elsewhere for the latest developments in connections between ethics and sustainability indexes.

In a general sense, *The Moral Austerity of Environmental Decision-Making* edited by John Martin Gillroy and Joe Bowersox and *Rationality and the Environment* by Bo Elling discuss the need to explicitly integrate ethics into environmental policy-making and explore strategies for doing so (Elling 2008, Gillroy and Bowersox 2002). These texts, however, focus on the need to explicitly integrate ethics into environmental policy-making and explore theories of doing so, also exploring theories of community participation in the process. They do not include concrete ethical examinations of indexes, or explore how community visions of sustainability can influence national sustainability assessments.

Though not trained as ethicists, John Peet and Harmut Bossel's "An Ethics-Based Systems Approach to Indicators of Sustainable Development" is one of the rare texts which explicitly articulates a theory about how ethics does and should influence sustainability indexes (Peet and Bossel 2000). Appropriately, they argue that ethics will always, often implicitly, influence indicator choice and advocate for explicitly attending to ethics during index construction. More problematically, they only justify their environmental ethical ideas to guide the development of sustainability indexes by the process of "adopt[ing] an ethic that makes most sense to us" (Peet and Bossel 2000: 224). In their cursory treatment of this subject, and given their focus on indicator theory rather than ethics, they simply do not have enough time to justify their approach or explore the meaning and implications of their ideals. Thus, it is not clear whether Peet and Bossel think everyone should use the ethic they articulate in their article or whether each group should use whatever ethic makes sense to it. More reflection is needed to understand how to facilitate the cooperation necessary for constructing and applying sustainability indexes in heterogeneous communities, nations, or the world without imperialistically imposing one ethic or thwarting all conversation as each group relies on its own ethics.

Bryan Norton's *Sustainability: A Philosophy of Adaptive Ecosystem Management* (2005) took one of the most significant steps toward bridging the gap between technical and ethical aspects of sustainability in general and with respect to monitoring progress toward sustainability in particular. He advocates for examining values in sustainability discussions, interdisciplinary conversation and participatory procedures for environmental decision-making, the mutual influence of facts and values, and the revision and adaptability of ideas and actions. Thus, throughout his work, he integrates the methods and content of the natural and social sciences with philosophical analysis as he explores how philosophy can improve environmental discourse, and as a side benefit, indexes. While groundbreaking, this focus means that Norton does not examine particular indexes or adequately discuss the complications that arise when constructing indexes. Additionally, he does not address issues important to the incorporation of ethics into sustainability indexes such as the identification and development of general principles which resonate with different ethical systems, or how normative visions of sustainability from a variety of scales (local, regional etc.)[3] can be aggregated into larger indexes (national, international etc.). Such issues are critical because, while local indexes are important, people also expect, develop, and use indexes to accompany national sustainability policies but do not want to give up the ethical priorities deeply rooted in their traditional, possibly local, cultures, philosophies, and religions to do so.

In contrast to Norton's focus on environmental discourse, social scientists such as S. López-Rideau et al. have worked to aid the development of local, community-based sustainability indicators (Lopez-Ridaura et al. 2002). López-Rideau et al. focus on traditional agricultural systems, describing a method generalized after working with over twenty local communities to develop sustainability indexes informed by their values. While their approach offers valuable insights into tested methods of community involvement in index creation, more work is needed to scale this process up past the scale of small local communities, since many sustainability indexes are used and useful at larger scales.

In sum, though some studies of ethics and sustainability indexes exist, this work is fragmented in that theories of bringing norms into the process are often disconnected from the analysis or development of actual indexes. Furthermore, most literature on sectorial-specific definitions of sustainability and sustainability ethics, whether about definitions or sustainability's connections to preexisting ethical systems, are separated from each other and from index development. While not every environmental or sustainability ethicist needs to focus on indexes, more integrative work is needed. Otherwise, critical questions of whose ethics should influence sustainability indexes, how diverse ethical traditions can influence sustainability indexes in a multicultural world, and how ethical and technical elements can and should relate in index construction, implementation, and evaluation are left relatively unexplored. These issues, and their integration, are critical to developing and implementing sustainability indexes that are technically robust and align with deeply rooted normative priorities, necessities of moving toward sustainability.

2.3.4 The environmental justice movement

As we have seen, justice is one of the recurring ethical commitments in the sustainability movement, so it is not surprising that one of the places where ethical and technical assessments have been more intimately and often explicitly linked in recent years has been in the environmental justice movement. It arose out of 1) the recognition by community members that their communities (often comprised of people of color or the poor) experienced degraded environments (and therefore degraded health and well-being) and less ability to meaningfully participate in decision-making about such issues than other people for no justifiable reason, and 2) the documentation of such trends by social scientists. Concern for environmental justice is often linked to sustainability because, as we have seen above, equity and justice, especially intra and intergenerational justice, are critical priorities of the theory and practice of the sustainability movement. Indeed, it is often argued that just life for all people is necessary to reach the social, economic, and environmental goals of sustainability (Agyeman 2005, Agyeman 2007, Agyeman et al. 2002, Agyeman et al. 2003, Agyeman and Warner 2002, Bullard 1994a, Geczi 2007, Hess and Winner 2007, O'Riordan 2005, Schlossberg and Zimmerman 2003). To demonstrate these connections this section examines definitions of environmental justice, their connections to ethical assumptions in sustainability definitions, and the ways technical and ethical elements are interwoven in the environmental justice movement.

Concern for the ways environmental issues affect subpopulations of society has existed for at least a century, as exemplified by initiatives focused on sanitation, public health, and workplace safety (Gottleib 1993). Grassroots activism about environmental degradation harmful to people who did not benefit from, choose, or know about the activities prompting the degradation, such as the efforts of Lois Gibbs and others at Love Canal in the 1970s and 1980s or that of Cesar Chavez and the United Farm Workers is now also seen as a part of the environmental justice movement (Chavez 1993, Cole and Foster 2001: 19–33, Figueroa 2002: 167, Gibbs 1999). Yet environmental justice came to be recognized as a distinct type of activism and subject of academic inquiry with the publication of landmark studies by Robert D. Bullard and the United Church of Christ's Commission on Racial Justice. They documented trends in the racial and economic make-up of communities in close proximity to toxic wastes: people of color and the poor in the United States disproportionately lived near toxic waste facilities (Bullard 1990, Commission for Racial Justice United Church of Christ 1987). The UCC found in its initial and follow-up studies that race was and is correlated with proximity to toxic waste facilities at a higher rate than economic factors (Commission for Racial Justice United Church of Christ 1987: 10, Goldman and Fitton 1994: 3, 5, United Church of Christ Justice and Witness Ministries 2007: 43).

Continuing in this vein, much early and contemporary environmental justice activism and research has focused on proximity to hazardous waste sites (Bullard 1994b, Checker 2002, Corburn 2005, Kraus 1993, Krieg 1998, Litt et al. 2002). But environmental justice initiatives have also examined who is harmed by invasive species and herbicide use (Chavez 1993, Norgaard 2007), urban heat island effects (Harlan et al. 2008: 187), asthma rates in cities, the cultural and health impacts of fishing in polluted waters (Corburn 2005), and the justice dimension of interaction between indigenous groups and more recent immigrants about the use of land (Checker 2002, Low and Gleeson 1998, Martin 2002). Environmental injustices have been documented all over the world with respect to a number of demographic groups including the poor, people of color in general or with respect to particular races or ethnicities (e.g. in America African Americans, Hispanics, and Native Americans), children, and women. As will be discussed in more detail in Chapter 6, a wide variety of technical methods including traditional statistical analyses and GIS studies of directly measured or survey data is used to document injustices. What these studies have in common is the repeated finding, both directly by community members and through the formal research of social scientists, that some groups of people experience greater environmental burdens or risks than others because they are members of particular racial, ethnic, economic, gender, or age groups, rather than for an ethically justified reason. Coupling these results; the experiences of environmental justice activists; and the civil rights, feminist, and mainstream environmental movements with the conviction that all humans have value and the right to meet their basic needs led to the birth of environmental justice studies (Cole and Foster 2001, Kraus 1993).

Documenting and protesting disparities are not, however, the only aspect of environmental justice; significant efforts have also been made to define types of environmental justice. Distributional and participatory justices, the most common terms, are defined as the inverse of environmental injustices. Distributional environmental injustices occur when demographic groups, most often people of color and the poor, experience effects of environmental degradation including illness and death and the loss of economic opportunities, ecosystems, and cultures at rates that are disproportionate to the benefits they receive from activities that cause the degradation. For example, native peoples who live in the Arctic and on small, low-lying islands are disproportionately affected by climate change: their traditional ways of life and often, their homes, are threatened by warming temperatures and rising sea levels though they have contributed little to climate change. On the other hand, people who have most benefitted from the rampant use of fossil fuels and other activities that contribute to climate change are and will be less affected by climate change. Distributional environmental justice, then, occurs when the benefits and burdens of human activity are experienced by the same population and all have the ability to access environmental goods and services to, at minimum, meet their basic needs. Not surprisingly then, discussions of distributive injustices usually focus on the distribution of environmental goods (e.g. clean air, safe drinking water, healthy food) and bads (e.g. toxic waste, smog, nuclear waste, noise, heat, and climate change). It can, however, also include the distribution of legal benefits and burdens regarding environmental issues (Gorovitz Robertson 2008: 537–39).

The just distribution of influence on and benefits of environmental decision-making is often classified as a separate type of environmental justice, participatory justice. This move ensures that participation in decision-making is not overshadowed by the study of the distribution of physical benefits and burdens. Participatory injustices occur when the ability of a group of people to participate in decision-making that affects them is inhibited by factors that are unrelated to the issues being discussed and are morally unjustified.

Barriers to participation may include laws which explicitly deny certain groups the ability to vote and/or participate in decision-making as well as political and social structures that effectively, even if unintentionally, prevent the ability of people to participate in decision-making and/or protest the decisions made (Figueroa and Mills 2001: 427–28, Harding 2007: 4–5, Ryall 2007). For example, public hearings which are not advertised sufficiently, are held at times in which local stakeholders are unlikely to be able to attend, or are advertised or conducted in languages in which the community is not fluent (e.g. highly technical language or official languages rather than the native languages of communities) can hinder the ability of people to contribute to decision-making that affects them, and result in injustice (Cole and Foster 2001: 7, 83–86, 89, 93–95, 109–12, Figueroa 2002: 169). Environmental injustices can also occur when decision-making processes only consult local stakeholders to placate them but do not actually factor their opinions and knowledge into the decision-making process. Additionally, a lack of access to information about environmental situations or the tools to understand or collect this information can contribute to participatory environmental injustices (Figueroa 2002: 174, Gibbs 1999: 89–90, Marc 2005). Conversely, participatory justice is believed to occur when people have the knowledge and ability to participate in decision-making and their input is taken seriously (Hunold and Young 1998).

The inability to participate in a meaningful way in decision-making that affects oneself is considered an injustice of its own by those who think that having some influence over one's life is a right. It is also deemed important because participatory injustices can lead to or exacerbate distributive injustices. After all, direct decision-making power or adequate representation in decision-making has long been recognized as the way to protect one's interests and ensure that a few who have decision-making power cannot favor themselves to the extent that the implications of their actions harm others.

While distributive and participatory environmental justice are defined against past and present injustices, the third, emerging type of environmental justice, restorative justice, focuses on future relationships of those who perpetrate and suffer from environmental injustice, now and/or in the past. It aims to ensure that relationships between these groups are not only restored to conditions before the injustice occurred but improved to an unimpaired state, in which community relationships will be strong and healthy and future injustices reduced. Restorative justice efforts are usually significant community endeavors in which parties recognize the scope and extent of the damage to ecosystems, communities, and individuals; make apologies where needed; and work to change future behaviors (Conrad 2011: 349–51, Figueroa, Personal Communication 2011, Figueroa and Waitt 2010: 140–42). Sometimes restoring environmental, economic, or cultural opportunities is a part of restorative justice (Dorsey 2009). Yet, direct restoration may not be possible when environmental or communal features are damaged beyond repair, as when people are killed by toxins or a sacred grove is destroyed. Thus, restorative justice may necessitate significant creative collaboration to recognize and deal with past injustices to improve future relationships.

All three types of environmental justice are connected to sustainability though this analysis will focus on the much more thoroughly developed connections between distributional and participatory environmental justice. For instance, the Brundtland Report's famous definition of sustainable development, "meet[ing] the needs of the present without compromising the ability of future generations to meet their own needs," includes both inter and intragenerational justice (World Commission on Environment and Development, 1987). Both of these forms of justice are typically interpreted as types of distributional justice as people assume that environmental resources should be distributed so that all can meet their needs. While the sustainability movement has primarily focused on the distribution of environmental *benefits* it also recognizes that environmental *burdens* have not been and are not being distributed equitably among people as it examines problems such as climate change which disproportionately affect people who have minimally contributed to it (Intergovernmental Panel on Climate Change 2007: 48–52, Stern 2006: 59).

Participatory justice is also featured significantly in the sustainability movement. At least since Agenda 21 and especially in local Agenda 21 initiatives, people are encouraged to involve local stakeholders in the process of defining visions of sustainability and making and implementing action plans to move toward it (Fernandez-Sanchez and Rodriguez-Lopez 2010: 1193, Holden 2011: 313). While sustainability experts may support stakeholder participation for pragmatic reasons (e.g. to ensure that new initiatives benefit from local knowledge and will be supported by locals and thus be more likely to be successful) many also support participation to ensure that the autonomy, rights, and values of stakeholders are respected. Indeed, participatory processes are increasingly encouraged within sustainability studies as policy-makers acknowledge the autonomy and rights of stakeholders, grassroots sustainability initiatives continue to grow, and more methods are developed to productively involve and share information with stakeholders (Elling 2008, Emmett and Desai 2010, Fraser et al. 2006, Norton 2005: 388–439, Reed et al. 2008).

Thus we see that participants in the environmental justice movement are often motivated by their experience and/or knowledge of disparities in environmental experiences correlated with particular demographic groups as corroborated by social scientific research and deemed unjust by ethical reflection. Here ethics and technical studies are intimately woven together, as is concern about the interaction of the environment, economy, and society, the three dimensions of sustainability. Yet, as will be described in more detail in Chapter 6, the social scientific studies of environmental justice are generally more focused on particular injustices or places rather than developing easily replicated indexes and there are few attempts to incorporate sophisticated methods of monitoring environmental justice into sustainability indexes despite the intellectual overlap between ideas of justice in the movements. Thus, the environmental justice movement, while resonant with the overall ideas of the sustainability movement, often remains distinct where indexes are concerned. Consequently, while the environmental justice movement to some degree counters the trend since 1992 of dividing technical and ethical aspects of sustainability, it does not attempt to answer the many questions about whether ethics can and should relate to sustainability indexes.

2.3.5 Rio+20

A similar pattern is found in *The Future We Want*, the outcome document of the 2012 UNCED known as Rio+20. Compared to the explicit separation of ethics and technical considerations in the original Earth Summit documents, *The Future We Want* integrates ethical commitments and technical assessments more thoroughly and takes steps toward overcoming the ethical limitations of earlier documents, though it still has a significant way to go, as much of the international community observed. But regardless of how the ethics of *The Future We Want* are assessed, one thing is clear: it continues and possibly increases the emphasis on monitoring progress toward sustainability but does not call attention to the need to integrate ethics into index development or develop methods by which to do so.

Rio+20 aimed to "secure renewed political commitment for sustainable development, assess the progress to date and the remaining gaps in the implementation of the outcomes of the major summits on sustainable development, and address new and emerging challenges" (United Nations 2012b). The conference had "two themes: (a) a green economy in the context of sustainable development and poverty eradication; and (b) the institutional framework for sustainable development" (United Nations 2012b) and placed special emphasis on the seven priority areas of "decent jobs, energy, sustainable cities, food security and sustainable agriculture, water, oceans and disaster readiness" (United Nations 2012b). By bringing together governmental representatives with nongovernmental organizations, Rio+20 aimed to:

> agree on a range of smart measures that can reduce poverty while promoting decent jobs, clean energy and a more sustainable and fair use of resources. Rio+20 is a chance to move away from business-as-usual and to act to end poverty, address environmental destruction and build a bridge to the future.
>
> (United Nations 2012b)

Thus, from its initial goals we see an emphasis on both economic and environmental concerns as well as ethical issues such as the "fair use of resources" or the quest for a good quality of life through decent jobs and ending poverty.

With respect to the particular ethical principles that are assumed in *The Future We Want*, we see a significant continuity with the ethical commitments of Agenda 21 as *The Future We Want* implicitly or explicitly supports norms of cooperation, responsibility, careful use, farsightedness, adaptability, adequate assessment of the situation, and justice. *The Future We Want* explicitly emphasizes cooperation and coordination of efforts in and between nations among governments, nongovernmental organizations, businesses, and civil society (United Nations 2012a: 54, 66, 189, 230)[4] about sustainability in general (United Nations 2012a: 54, 55, 63, 124) and with respect to specific elements of sustainability including agriculture (United Nations 2012a: 114), economics (United Nations 2012a: 19, 58, 209), communication technology (United Nations 2012a: 44, 65), water treatment and use (United Nations 2012a: 124), the human rights of migrants (United Nations 2012a: 157), "disaster risk reduction strategies" (United Nations 2012a: 187), biodiversity preservation (United Nations 2012a: 202), and improving education (United Nations 2012a: 229) among other issues (United Nations 2012a: 54, 66, 189, 230, 260).

Human responsibility for environmental degradation and for taking action to move toward sustainable development is explicitly stated and implicitly assumed throughout *The Future We Want* more than in the Earth Summit documents as the word is used more often and because the later document emphasizes ethical duties more throughout. This is particularly seen in its emphasis on the careful use of resources and the treatment and disposal of wastes (United Nations 2012a: 202, 220, 225, 227, 252) and in exhortations to carefully utilize all sorts of natural resources whether biodiversity, water, air, mountains, or land (United Nations 2012a: 135, 141, 159–78, 197–212). Significantly, however, *The Future We Want* does not emphasize the fact that some nations, particularly industrialized countries such as the United States, have been much more responsible for many types of environmental degradation. Similarly, the document embodies the principle of farsightedness insofar as it emphasizes that the use of goods and services can have far-reaching spatial and temporal effects if pollution is directly shipped or indirectly travels elsewhere (United Nations 2012a: 225), if activities contribute to climate change (United Nations 2012a: 25, 111, 128, 166, 178, 190–92), or if resources used in one country come from another and their extraction damages the other country, though it only stresses that some countries (especially small island and African nations) are more at risk from environmental destruction such as climate change without explicitly naming the other nations who are to blame for such situations.

Adaptability is also quite prevalent in *The Future We Want* as the document affirms that due to the different cultural, economic, and environmental situations of various countries and regions, people will have different views of sustainable development and different pathways to move toward their targets (United Nations 2012a: 32, 34, 37, 58, 59, 63, 133, 178, 186) that currently exist or may exist in the future as situations change (United Nations 2012a: 75). For example, when discussing agriculture, it "resolve[s] to improve access to information, technical knowledge and know-how, including new information and communications technologies that empower farmers, fisherfolk and foresters to choose among diverse methods of achieving sustainable agricultural production" (United Nations 2012a: 114). This approach is key to the document's ethical and technical plans as it allows signatories to acknowledge the differences between their constituencies while envisioning a sustainable future.

According to *The Future We Want*, an action plan for sustainable development also includes a firm commitment to the best possible knowledge about ecosystems, economies, and societies in order to make informed decisions and work within the limits of existing and future knowledge (United Nations 2012a: 18, 34, 47, 62, 63, 66, 136, 172, 203, 208, 220, 239, 274). This commitment to an adequate assessment of the situation is seen in its multiple exhortations to gain the best and most thorough knowledge possible. While using and increasing the availability of reliable, accurate data has long been an emphasis of international discussions of sustainable development, *The Future We Want* makes two shifts when compared to Agenda 21. First, it not only emphasizes data collection in general, but also in multiple places emphasizes that gender-specific data is needed since women often bear the brunt of environmental devastation and will often be key to sustainable development as they are firsthand resources for agricultural sustainability because they are such a large proportion of the world's farmers and food producers; because they choose fuels and prepare foods; and because their empowerment is often correlated with increased quality of life not only for women but for all people (United Nations 2012a: 136, 191, 239). Second, it more significantly emphasizes the need for data collection about all three dimensions of sustainable development, particularly the social, which tended and still tends to be neglected in measures of progress toward sustainable development and in discussions of them as found in Agenda 21. For example, *The Future We Want* acknowledges the importance of "physical, mental, and social well-being," and advocates collecting data to ensure that "the highest attainable standard of physical and mental health" can be obtained, an advance when compared to the emphasis on physical health in earlier documents (United Nations 2012a: 138). Similarly, *The Future We Want* recognizes that communities can contribute significantly to the way progress is measured as it advocates for "the further development and implementation of scientifically based, sound and socially inclusive methods and indicators for monitoring and assessing the extent of desertification, land degradation and drought" (United Nations 2012a: 208).

These changes to the monitoring of data, specifically to widen the types of data that are being used, signal and speak to the fact that *The Future We Want* has an increased and more fully developed commitment to justice, social issues, and ethics in general compared to Agenda 21 and the Rio Declaration. Specifically, *The Future We Want* has a wider recognition of things to have justice about – social conditions as well as physical and economic conditions, as described above, and the increased number and types of groups that are recognized as significant and in need of justice.

While Agenda 21 emphasized the major groups of children, youth, indigenous people, nongovernmental organizations, local authorities, workers and unions, business and industry, the scientific and technical communities, and farmers (Robinson et al. 1993: 492–545) *The Future We Want* also mentions ethnic and cultural groups, disabled people, the elderly, and any poor and vulnerable people, indicating that it begins to recognize that more groups can face distributive and participatory injustices. For instance, *The Future We Want* particularly highlights the plight of Africans and those in small island nations as those who are affected by environmental degradation, including that from climate change, though they do not necessarily, and indeed, often are not the most significant instigators of these problems (United Nations 2012a: 16, 17, 32–33, 35, 165, 174–80, 183–84, 205, 265). It also places a more significant emphasis on women than previous documents.

We also see the importance of justice in *The Future We Want* as the document goes farther in its discussion of the fair use of environmental resources identified by indigenous groups. It not only identifies the knowledge of indigenous groups as a resource for other people but also explicitly identifies the cultural heritage and life of the indigenous people themselves as something valuable, something that Agenda 21 placed less emphasis on (United Nations 2012a: 43, 49, 58, 71, 175, 197, 211, 229). With this increased attention to justice, it takes steps toward addressing some of the ethical critiques of the Earth Summit documents, particularly those about the need to attend to the disparities between the environmental experiences of various groups.

These advances are significant in the advocacy for and hopefully, the achievement of better justice as sustainable development action plans are put into place and evaluated. Unfortunately, however, *The Future We Want* does not emphasize the collection of disaggregated data to track potential injustices, say with respect to ethnic groups, aside from gender differences. Since measurable data so often is taken as a demarcation that a problem is "real" or "significant" and is often required for transparent decision-making in democracies or shareholder owned businesses, it is unlikely that this focus on justice for various groups will be upheld without more concern for the data as well. Thus, while *The Future We Want* is a step in the direction of justice, more steps are needed to attain the desired future.

The Future We Want also works to overcome critiques of the Earth Summit documents by recognizing the intrinsic value of nonhuman biota (United Nations 2012a: 39, 56, 163, 197, 202, 228) and acknowledging that some signatories emphasize the rights of nature or the idea of "Mother Earth" (United Nations 2012a: 39). More important for this argument, however, is the fact that *The Future We Want* better integrates ethical and technical concerns. This is probably best illustrated in its increased attention to all three dimensions of sustainable development and its broader consideration of what a good quality of life entails. While the Earth Summit documents, particularly Agenda 21, focused upon economic and strictly environmental aspects of sustainable development – financial valuing of natural systems, air and water quality, and so on – *The Future We Want* gives increased attention to the social components of sustainability. We have already seen examples of this when it acknowledges the importance of mental health and recognizes that cultural preservation and identity are significant and should be valued in the quest for sustainable development, particularly among "indigenous peoples and their communities, other local and traditional communities and ethnic minorities" (United Nations 2012a: 58, 156). Society is also emphasized when *The Future We Want* recognizes that "social policies are vital to promoting sustainable development" (United Nations 2012a: 63) and that "normsetting" for sustainable development should include widespread participation of people from around the world including from developing countries (United Nations 2012a: 92). It also weaves this multifaceted concern throughout the document's analysis of specific issues including water and sanitation services, land management, and decision-making, including the collection of data about sustainable development (United Nations 2012a: 63, 119, 120, 207). In these examples and others throughout the document, we see that *The Future We Want* highlights a wide range of priorities beyond economic strength and environmental protection to benefit society (United Nations 2012a: 125, 132, 134). While this broad range of commitments is not unique to *The Future We Want* among international sustainable development plans, its broader attention to social matters is more detailed and focused on cultural preservation and quality of life than basic physical and economic needs, as had been emphasized earlier. This broadening of considerations indicates a shift to acknowledge societal and ethical considerations as significant, to recognizing that economic and environmental conditions are not the only important concerns when working for sustainable development.

Despite these moves to an increased rhetoric of ethics, many nongovernmental organizations, pundits, and activists have seriously critiqued the Rio +20 conference and *The Future We Want* (Ford 2012, Harvey 2012, Hawley 2012, Margolis 2012, Narain 2012, Women in Europe for a Common Future 2012). Critiques fall into three main categories. First, that the conference and outcome document did not achieve what they were supposed to: set clear targets for achieving sustainability. Critics claim that without a concrete action plan to combat climate change and other anthropogenic environmental problems, the conference was just blowing more hot air. Second, that the process of holding the conference and crafting *The Future We Want* was not sufficiently open to participation by the world community and that the input of nongovernmental organizations was not taken seriously. Third, that

56 *Sustainability*

The Future We Want did not take problems such as climate change and overpopulation seriously enough. For instance, as noted above, the document does not significantly discuss who is responsible for global environmental degradation or other moves away from sustainable development, nor who has failed to take sufficient action to address such problems in the last twenty years. Given these shortcomings, the rhetorical changes in *The Future We Want* as compared to the 1992 Earth Summit documents are seen as small concessions to placate protestors rather than making the significant commitments necessary to progress toward sustainability.

Beyond these critiques, however, is a problem as much if not more significant for this project. Namely, while *The Future We Want* heavily emphasizes the need to track progress toward sustainable development, and generally advocates participatory processes to do so, it does not spell out how social and ethical concerns are to be integrated into indexes or adequately illustrate such a process by example. Thus, while *The Future We Want* indicates at least a surface level acceptance of the importance of relating ethical and technical concerns with respect to sustainability initiatives, it leaves the difficult questions of how to link them during index development and use unexplored, and indeed, unacknowledged.

2.4 Implications

This pattern of focusing on either normative or technical aspects of sustainability has proliferated since the movement exploded after the Rio Summit. Disciplinary training which emphasizes methods of economic, biological, or ethical analyses contributes to this trend as does the need to simplify the complex problem of global sustainability in order to study it. Yet, as illustrated above, from its earliest precursors to the present day, sustainability has encompassed both what it is possible to sustain technically under a given set of circumstances and what people normatively desire to sustain, even if one aspect is emphasized at any one time. Certainly, there has been recognition of the need to link these aspects of sustainability, as demonstrated in the coupling of the Rio Declaration and Agenda 21 as well as the integrative work of both Daly and Cobb and Norton. Yet, this recognition has not adequately played out in the sustainability movement. Too often technical and normative assessments take the other as a starting point and exhibit little to no interest in further contact. This bifurcation belies the necessary intermingling of technical and normative assessments in general; the long history of linking them in sustainability studies, even if implicitly, especially in the study of sustainability indexes; and leaves a number of questions unexplored. How can norms significantly influence sustainability indexes while preserving the advantages of quantitative or qualitative progress toward sustainability? Whose ethics should influence sustainability indexes? If the ethics of people who will use and be affected by the index, what happens when an index is to be used in an ethically heterogeneous setting?

Furthermore, how can local ethical priorities, which may vary from place to place, influence statewide, national, or international indexes? Leaving these questions unaddressed hinders progress toward sustainability, for sustainability goals, as represented in indexes that are not normatively acceptable, will not be implemented (or not for long) while goals that are not technically feasible will not be reached. Thus, integrating these elements is necessary to make progress toward sustainability. To enable sustainability studies in general and indexes in particular to integrate these critical aspects of sustainability, Chapters 3 and 4 will develop a method of index construction and an ethical method to address these issues.

Notes

1 Portions of Section 2.2.3 were adapted from the author's article, Fredericks, S. E. (2012). Agenda 21. In W. Jenkins and W. Bauman (eds.), *Berkshire Encyclopedia of Sustainability* (Vol. 6: Measurements, Indicators, and Research Methods for Sustainability). Great Barrington, MA: Berkshire Publishing Co. Adapted with permission from Berkshire Publishing Group.
2 A few scholars have begun to articulate theories of how broad principles can put bounds on acceptable decision-making even though the principles can be specified in different, possibly contradictory ways (Neville, 2001c; Norton, 2005; Waltzer, 1994).
3 Norton does discuss scale, but he studies the scale at which a particular ethical principle applies, not the scale of the communities which adhere to possibly different ethical traditions.
4 References to *The Future We Want* include paragraph numbers rather than page numbers.

3 Index theory[1]

Typically, when individuals or societies want to achieve sustainability they do not just want to develop an ethical theory for sustainability or act in ways that they think will be sustainable. They also desire a method to assess their progress toward sustainability (Faucheux and O'Connor 1998: 4, 10, Levett 1998: 301, Mitra 2003: 30, Obst 2000: 8). Indicators and indexes enable such assessments and are often requested by policy-makers to help guide their decision-making (Astleithner and Hamedinger 2003: 627). So far, economists, ecologists, and experts on quality of life have been the key players in index development. Thus, it is not surprising that the technical aspects of sustainability have been emphasized over its normative aspects in index development. Some experts, however, have recognized the need for sustained ethical analysis of indicators given the many moral aspects of sustainability and indeed, ask for assistance from ethicists (Dahl 1997: 82). Now is a perfect time for such work: theories of index development are fairly well developed and some indexes exist yet they can still be easily revised since they have not yet been widely implemented.

To enable such work, two prior moves regarding indexes are necessary. First, the technical aspects of index theory, examined in Sections 3.1–3.3, must be understood to be integrated with ethics. Second, the existing and possible roles of ethics in index development and use must be identified as in Sections 3.4–3.5. Specifically, Section 3.1 outlines definitions of and uses for indicators and indexes. Section 3.2 explores frameworks for sustainability index development. Section 3.3 notes that index developers working within these frameworks face a number of tensions to be balanced during index construction including the desire for comprehensiveness and manageability, the need for technical rigor and accessibility, and the differences between ideal and actual data. Section 3.4 notes that ethics already influence the process of index formation, though such influence is often unintentional. Finally, Section 3.5 suggests ways to evaluate existing indexes and develop new methods of index development more explicitly attuned to ethics.

3.1 Definitions of indexes and indicators

Indicators and indexes are used in a variety of fields including economics, sociology, medicine, the environmental sciences, and policy-making in order to simply represent something about a complex system such as an economy, society, the human body, or an ecosystem. Indicators and indexes are symbols in that they represent an aspect of a system to someone and are typically mathematical representations of functions of data about a system. The most thorough indexes will attempt to encompass the entire system including its components, overall characteristics, and the connections between them (Gallopin 1997: 18–19, van den Begh and Verbruggen 1999: 62). System changes can be tracked over time by periodically recording and comparing index outcomes (van den Begh and Verbruggen 1999: 62).

A common index is the UV index published daily in many newspapers and weather websites. It is a quantitative index rating the ultraviolet light danger to human skin using a color-number system; low numbers and cool colors (blue/green) are safer than warm colors (yellow, orange, and red) and numbers closer to 11 which correspond to increasingly dangerous rays for a particular time and place. In America, the National Oceanic and Atmospheric Administration (NOAA) calculates the UV index from data on elevation, predicted cloud cover, incident UV radiation, and ozone levels for a particular time and location (US Environmental Protection Agency 2006). Given the complexity of this data, the UV index is a convenient indicator of sunburn risk compared to collecting data and calculating the risk on one's own, especially if traveling to an unfamiliar place. Thus, the UV index illustrates one of the most important features of indexes: they simplify a large amount of information about a complex system to inform decision-making.

While "indicator" and "index" are sometimes used interchangeably, "indicator" usually connotes a piece of data (i.e. elevation, cloud cover) while "value of the index" or "index" refers to the output of a mathematical function aggregating indicators (i.e. a number 1–11). "Index" can also refer to the whole function itself (i.e. the UV index) (Gallopin 1997: 15, Ott 1978: 8). Sometimes indexes are made of several components called subindexes or subindicators that are themselves functions of indicators. For example, sustainability indexes may be based upon a three-dimensional framework in which progress toward sustainability is reached through simultaneous economic, environmental, and social development. In this case, a function representing a dimension of sustainability, say GDP for the economic dimension, is a subindex even though it may be called an index when used alone. Thus, the classification of a particular variable as a subindex as opposed to an index or indicator depends on the context in which it is discussed.

Indexes can assess current conditions, serve as warnings, predict future trends, and possibly suggest when future studies are needed to understand a phenomenon. Most importantly for this study, indexes can compare situations to track progress, or lack thereof, toward policy targets (Carlisle 1972: 28–31, Failing and Gregory 2003: 122, 124, Gallopin 1997: 15, Ott 1978: 4). It is generally recommended that policies and indexes that monitor progress toward their goals be constructed at the same time to ensure that indexes align with critical features of the policy (Carlisle 1972: 28–31, Levett 1998: 191, 293,

60 *Index theory*

Obst 2000: 14, van den Begh and Verbruggen 1999: 62). In reality, indexes may or may not be developed along with policies. Preexisting indexes are often used with new policies to maintain continuity with existing data or to save the time and money required to collect data and construct indexes. Academics may develop new indexes to test new monitoring methods and/or call attention to previously understudied elements of the system in question before policies are changed. For example the three-dimensional index of Sustainable Energy Development and Robert Prescott-Allen's Wellbeing of Nations, indexes assessed in Chapter 5, were developed to assess a country's progress toward sustainability apart from any one policy (Davidsdottir et al. 2007, Prescott-Allen 2001). Whether or not indexes are developed separately from the policy-making progress, the significant growth in the number of sustainable indexes is largely attributable to the international focus on sustainable development after the 1992 Earth Summit, particularly Agenda 21's call to monitor progress toward its goals.

3.2 Frameworks for sustainability indexes

Indeed, the United Nations created the Commission of Sustainable Development (CSD) after the Earth Summit to "monitor the progress that has been made on the way toward a sustainable future" (Moldan et al. 1997: 1). Yet efforts to develop sustainability indexes have faced significant challenges as sustainability and sustainable development are notoriously difficult terms to define and can refer to many complex, dynamic systems at a variety of spatial and temporal scales. Thus, shortly after the Earth Summit frameworks were developed by the CSD and other groups to delineate the scope of sustainability indexes and set standards for their development (Becker 2004: 200–4, 206–8, Bosivert et al. 1998: 100–1, 106–7, 110, Gallopin 1997, Moldan et al. 1997: 1).

Frameworks aim to categorize indicators, ensure that index developers consider all relevant information, and aid the interpretation of index results (Gallopin 1997: 21). To accomplish these goals, frameworks often divide sustainability into categories: according to media (land, water, air, and biota), according to economic sectors (transportation, industry, urbanization, agriculture), or according to particular policies (Gallopin 1997: 21–22). They may also focus on incorporating the environment into national accounting systems by using money as the unit of measurement. Alternatively, they may focus entirely on environmental statistics (Gallopin 1997: 22, Holub et al. 1999: 236, Obst 2000: 10–13). Two sorts of frameworks, however, are most prevalent in indicators of sustainable development: dividing sustainability into economic, ecological, and social dimensions and the Pressure–State–Response framework and its variants (Gallopin 1997: 22, Moldan et al. 1997: 1, Obst 2000: 7).

Breaking sustainability into economic, environmental, and social dimensions is quite popular because this classification schema is thought to encompass all important elements of sustainability while ensuring that each receives attention (Faucheux and O'Connor 1998: 3, Moldan et al. 1997: 1, Obst 2000: 7). The environment is often named a dimension of sustainability because it allows human society to survive, but is generally considered separate from society. Additionally, some argue that it is valuable in and of itself and therefore should be sustained regardless of its connection to humans. The environmental dimension often includes assessments of ecosystem integrity and land, water and air quality including, for example, measures of emitted pollutants, deforested land, or threatened species. Yet, sustainability advocates who developed the three-dimensional framework do not merely desire that ecosystems or life in general is able to be sustained; they also aim to sustain aspects of human society. They demonstrate this commitment by naming the social dimension, which typically focuses on basic quality of life issues such as life expectancy, educational rates, and peace. In contrast, the economic dimension aims to capture the health of the economy as a whole and is usually measured by GDP though savings, unemployment rates, deficits, and other factors may also be considered. Though the economy arises out of human society, it is generally considered a separate dimension since economic resources are believed necessary to achieve social and environmental sustainability. This separation is partially an historic artifact: economic measures of sustainability compromised the first explicit sustainability indexes. Only later were other aspects added as it was recognized that economic strength did not fully capture societal well-being.

By separating the three dimensions when identifying and selecting sustainability indicators, index developers aim to recognize all dimensions and ensure that if one dimension is emphasized at the expense of another, such prioritization is acknowledged. Historically, the three-dimensional framework drew attention to the often neglected social dimension. Neoclassical economics do not heavily value justice or quality of life and the significant amount of research on environmental sustainability has overshadowed studies of social sustainability (Faucheux and O'Connor 1998: 13).

The advantages of separating sustainability into three dimensions can, however, turn into limitations if taken too far. The three-dimensional concept can force false divisions between indicators of sustainability when they must be categorized in one dimension or another to follow the framework. This raises questions about the degree to which indicators overlap and how to avoid double counting if overlapping indicators are included in an index. The separation of dimensions can also deemphasize the deep connections between them. The division between the economic and social dimensions often feels most problematic for as we know, the economy is a social artifact. It owes both its basic existence and its character to humanity: without us there would be no economy. Thus, the economy is primarily, if not entirely, a means to an end – the flourishing of human society – even if it often appears to be or is valued as an end in and of itself (Bosivert et al. 1998: 100, Faucheux and O'Connor 1998: 3–4, Levett 1998: 295–296). Recognizing and guarding against these potential limitations is necessary to ensure that they do not overshadow the advantage of the dimensional framework – it promotes the recognition of all three dimensions.

In contrast to the dimensional comprehensiveness of the three-dimensional framework, the Pressure–State–Response (PSR) framework and its variants, including the Driving-Force–State–Response (DSR) framework, classify subindicators by the function they monitor. Driving-Forces are "human activities, processes and patterns" that influence sustainable development either negatively or positively (Forstner 1997: 220), a revision of the pressure category which only included negative influences. Driving-Forces can, for instance, include the amount of nuclear waste generated or groundwater consumed per year, the amount of energy consumed per year or per GDP, or the amount of restored wildlife habitats per year. State indicators are those that measure the quality or quantity of the state of a component of the system. State indicators for the environmental dimension, for example, may record the population size of a species or the concentration of an air pollutant (Bosivert et al. 1998: 109, Gallopin 1997: 22, O'Connor 1998: 32). Response indicators include the actions proposed by or actually undertaken by society to respond to the environment as well as measures of the willingness or effectiveness of society's responses (Bosivert et al. 1998: 109, Forstner 1997: 220, Gallopin 1997: 22, O'Connor 1998: 32). For instance, records of laws enacted or law enforcement rates can be Response indicators.

While the general categories of Driving-Force, State, and Response indicators may seem to indicate that a causal chain links them such that a driving force changes the state of a system, provoking a response among humans, most authors are careful to point out that there is no implied or necessary causality between the individual indicators in each category. Some such causality may exist, say between the energy consumption per capita and the concentration of carbon dioxide in the atmosphere. However, indicators in the functional categories are not always linked, nor should assumptions about their connections be made without careful study. For example, carbon dioxide is emitted through many activities besides energy production and not all forms of energy production emit carbon (Mortensen 1997: 51). Thus, even though connections between these two indicators may exist, any particular Driving-Force indicator is not necessarily the only influence on any particular State indicator. Indeed, there is no guarantee that even a partial cause of one of the State indicators will be included in any set of Driving-Forces. Thus, the Organization for Economic Co-operation and Development (OECD) cautions people against assuming that connections between indicators are linear or mechanistic, or indeed, that they exist at all (Gallopin 1997: 23–24).

The DSR framework can, however, help index developers determine the type of indicators best suited to their goals. For instance, policies designed to curb industry emissions would be most effectively monitored by Driving-Force indicators whereas an overall assessment of environmental health before, during, and after decades of efforts to curb acid rain may be best monitored by State indicators of forests or lakes during these time periods. If data is not available for the preferred types of indicators then index developers must carefully justify the links between the indicators they have data for and those they wish to learn about.

As the three-dimensional and DSR frameworks target different aspects of sustainability indicators, they may be employed simultaneously to identify indicators that fulfill all nine possible combinations of the two frameworks (e.g. Driving-Force economic indicators or State environmental indicators). So far, most attempts to link the DSR and three-dimensional frameworks have resulted in lengthy lists of indicators (United Nations 2001: 312–313). Composite indexes that compile indicators to summarize the movement toward sustainability are still relatively theoretical. Of those that have been developed, many try to focus on a portion of sustainability to make their task manageable. Some center on a particular segment of the functional framework – the Driving-Forces or the States – while others focus on a particular aspect of sustainability such as agricultural systems, human health, ecosystem integrity, or the impacts of energy use (Failing and Gregory 2003: 127, Holub et al. 1999, Hueting and Reijnders 2004: 258, van den Begh and Verbruggen 1999: 63).

3.3 Balancing polarities in index development

While the frameworks discussed above can be instrumental in determining the scope of a new index, they are not enough to ensure that sustainability indexes are functional. To be meaningfully implementable by laypeople, indexes must be relatively easy (and cheap!) to figure, comprehensible to the average person, scientifically sound, and nuanced enough to discriminate between policy options (Failing and Gregory 2003: 128, Hezri and Hasan 2004: 288, Levett 1998: 191, van den Begh and Verbruggen 1999: 62). Constructing a good index is not an easy process for each one is "always a compromise" (Bakkes 1997: 379) between three sets of polarities: 1) manageability and comprehensiveness, 2) technical rigor and accessibility, and 3) ideal and actual data. As benefits and drawbacks exist for points between these polarities, and the optimal balance of these polarities is heavily dependent on the situation in which and for which the index is constructed, no ideal combination will emerge. Thus, index developers must be aware of these tradeoffs to construct a suitable index for a particular situation.

Most simply, one can monitor the most important aspect of a system's progress toward or away from sustainability in one indicator. For example, given the threat of climate change to global sustainability and the fact that carbon emissions are its most significant cause, some use carbon emissions as *the* indicator of sustainability. Such single indicator indexes are relatively easy to measure and understand.

Yet single-indicator indexes face substantial challenges. Identifying the one most significant indicator can be difficult given the number of diverse criteria by which sustainability is judged. Prioritizing one indicator will necessarily neglect others and may limit knowledge of sustainability. For example, methane's contribution to climate change is ignored when carbon dioxide is taken as its sole indicator. Similarly, the one problem, one indicator approach to monitoring sustainability will draw attention away from other critical sustainability issues such as toxins.

To avoid the limits of extreme simplicity, some sustainability experts identify a list of indicators to comprehensively cover the central aspects of sustainability. For example, Eurostat, the statistical branch of the European Union, has identified and tracked more than one hundred indicators (Eurostat 2012d). Yet comprehensiveness can pose its own problems. Amassing all of the data to compute the values of extensive lists of indicators just once can be quite time-consuming and expensive, to do so year after year would be even more difficult. The most technically prized indicator set will not be implemented (or not for long) if it is not viewed as cost-effective. Additionally, when so many indicators are presented individually in a table the abundant information can easily overwhelm users. Then they may focus on just a few indicators that they personally understand or find more important even if others disagree with their valuation or other indicators have a more statistically significant impact on the system in question. Eurostat identifies eleven "headline indicators," the most significant, for policy-makers to focus upon, but even eleven discrete data points may confound an overall assessment of sustainability.

Additionally, technical experts and policy-makers may have different priorities on the scale of manageability and comprehensiveness. All too often technical experts hesitate to endorse the most manageable indexes because they know that many nuances of the data and their relationships are lost in such simplifications. Sometimes scientists would rather let the data speak for themselves than interpret it, not acknowledging that data always involves interpretation. On the other hand, policy-makers often know that they need technical guidance and would prefer clear, definitive answers: the overall air quality is getting better (or worse) (Ott 1978: 6). Consequently, indicator theorists will need to balance the competing impulses for manageability and comprehensiveness as they strive to create indexes or indicator sets that acknowledge the intricacy and uncertainty of complex systems yet produce strong results.

As developers of indexes strive to mediate the desire for comprehensiveness with the need for manageability many seek a technological fix – a composite index that combines the most important elements of a system to yield a single scalar or vector output. However, such aggregate indexes, even more than simple indicators, require the balance of technical rigor and a method of aggregation accessible to the general public.

If policy-makers and the general public fully understood an index and its results, they would be less likely to misapply the results of the index, as often happens, say when GDP is used an overall index of societal well-being.

To achieve accessibility, data collection, analysis and aggregation methods should be "clear, transparent and standard" (Gallopin 1997, 25). Using existing data sets will also limit political, social, economic, and temporal challenges to data collection. Even if meeting these standards for accessibility is possible, it is likely that average citizens and policy-makers are not interested enough to learn the subtleties of an index that ensures that it meets the technical criteria important to experts. Rather, policy-makers desire simplified information to guide their actions.

To meet these divergent goals, indexes are often constructed and explained through a hierarchical process involving increasingly less detail. The first stage, that of data collection and aggregation into indicators then subindexes and finally an index, is generally quite detailed. The second stage, that of the final product, which may be a number, graph, or other symbol, summarizes the knowledge about the system into a clear, direct result as desired by policy-makers; technical details are available for disaggregation and study by policy-makers. For instance, the 2012 Environmental Performance Index is an aggregated index comprised of ten subindexes corresponding to individual policy categories, each of which is comprised of one to four indicators (Emerson et al. 2012a). While experts may be interested in all of these subindexes and indicators, policy-makers will probably focus on one or two components or on the overall indexes. A similar process of aggregation is found in the narrative descriptions of indexes: one may read the details of data collection and aggregation in full methodological reports, or a brief summary of aims and outcomes in a brief interpretive statement for policy-makers. These dual approaches allow indexes to meet both expert and lay expectations.

The danger of this approach, however, is that decision-makers will not understand or recognize the assumptions of the index or its limitations, conditions that provide a fertile ground for misinterpretation or misuse. Employing GDP as an indicator of human well-being instead of just an economic indicator is one of the most famous overextensions of an indicator. To avoid this problem, index developers may recommend that indexes are specifically created to align with the objectives of a specific policy (Hezri and Dovers 2006: 319, Holden 2011).

Yet when index developers seek to develop new indexes they face another challenge: the need to balance the desire for ideal data sets with the advantages of using data that already exists. While index constructors may yearn for data directly related to their topic, limited budgets may intervene. Thus, either index developers must fight for the adoption of their index and the implementation of new data collection programs or they must use preexisting data as proxies for desired data. This tradeoff is immensely important because even if an index is perfect in theory, it will never be used if it relies on data that is immeasurable in theory or is too expensive or technologically or politically difficult to collect (Council 2000: 3). On the other hand, while proxies can be effective, it can be difficult to establish a direct link between a proxy data set and the desired data especially since proxies are used when the desired data is not consistently measured, if it ever has been. Using existing data sets can also bias indexes for they may be based on data that is available but not ultimately of premier importance for the system, or neglect important but unmeasured elements of the system (Failing and Gregory 2003: 129, Obst 2000: 15). Existing data favors industrialized nations and economic over environmental and environmental over social data such that using existing data sets emphasizes the economies of industrialized nations though they are but one influence on the system in question and maybe not the most important one (Burger et al. 2010, Gallopin 1997: 25, Levett 1998: 191, Pearsall 2010: 875). Data also tends to focus on the state of the environment itself (the number of species, amount of pollutants etc.) rather than how the environment and the human population relate. Such data, when incorporated into indexes, can lead to oversimplified assumptions of a dichotomy between humans and nature or the idea that anything good for humanity or ecosystems is good for the other (Azar et al. 1996a). Given these limitations, index developers should recognize and alert index users to the limitations of existing data.

Many index advocates argue that reliable, accessible, appropriate index data is essential to teach people about the system in question and promote good decision-making (Council 2000, Hezri and Dovers 2006: 358, Hezri and Hasan 2004: 288, Inhaber 1976: 4, 162–63). Thus, they believe that indexes are the key to proper, informed decision-making. Understanding a situation is helpful to making effective decisions, but values also implicitly influence the development, use and evaluation of indexes (Failing and Gregory 2003: 129, Hezri and Hasan 2004: 288, Levett 1998: 292–293, Moldan 1997: 99, Ott 1978: 3, Vatn 2005). Thus even if frameworks and guidelines are followed and a point along the spectra of 1) comprehensiveness and manageability, 2) technical rigor and accessibility, and 3) ideal and actual data is justified, index developers still need to attend to the normative concerns of the community as they build an index.

3.4 Normative priorities and index development

Though the influence of norms on index formation is understudied compared to the spectra in the previous section, normative priorities, both ethical principles and cultural values, influence the entire process of the development and use of sustainability indexes. Particularly, values inform 1) sustainability definitions which influence indexes, 2) the decision to develop an index, 3) the selection of methods of index development, 4) the actual process of index development, and 5) the use of indexes.

While ethical principles are the primary type of norm addressed in this project because they represent a community's ideals about their actions, the many values a community may have are also critical influences on sustainability indexes. For instance, communities may value education, freedom of speech, the observance of certain holidays, or caring for elderly relatives at home until the end of their life. Values may also concern the types of knowledge respected in the community, for example modern scientific knowledge, religious experience, or traditional ecological knowledge. While these values may be explicitly articulated they are often implicitly expressed in the community's actions, stories, and policies. Regardless of their mode of expression, they are normative because they are deemed the normal or right way to understand the world and shape all aspects of community life. Consequently, they guide the articulation and application of ethical principles and, along with ethical principles may influence index construction in five general ways.

First, as Chapter 2 demonstrated, normative dimensions of sustainability pervade its many definitions, even if they are explicitly focused on technical matters. For instance, ensuring that present and future generations have the ability to meet their own needs (justice) and taking responsibility for their actions are common ethical priorities of the sustainability movement. Grounding these ethical principles are community values about life, one's descendants, human society, and the rest of the world. Insofar as these norms are embedded in definitions and discussions of sustainability, which in turn influence indexes, they may influence indexes. Indeed, communities using indexes may expect such interconnection so indexes evaluate progress toward their vision of sustainability.

Norms also influence sustainability indexes since deciding to use indexes to monitor progress toward sustainability is a value-laden decision. It highlights the priority placed on quantifiable information, often on short-term monetary assessments, voters' desire to know how the fiscal and societal impact of their leaders' decisions, and the assumption that such knowledge should be available.

Third, the choice of participants in index development, whether scholars, lay people from constituencies impacted by the index's use, or a combination thereof, is shaped by social values. Public participation is increasingly seen as critical to identify index goals, levels of risk a community is willing to accept, and how indexes and policy interrelate (Astleithner and Hamedinger 2003, Becker 2004: 201, Dahl 1997: 81). Ideally, involving a diverse group in index development will enable the values and knowledge of the community to be utilized in the index, and the values and therefore the rights of all affected people to be recognized. Bryan Norton argues that such a decision is a normative move insofar as "The community has placed a value on achieving cooperative solutions to shared problems" (Norton 2005: 366). Norton himself favors such approaches because he thinks that a variety of participants helps guard against bias, and can come as close to objectivity as possible (Norton 2005: 348–86). While public participation is important, wide-scale agreement that a particular course of action is normatively desirable does not, of course, mean it is in fact technically sustainable. Thus, careful attention to index construction by both experts and lay people is needed to ensure the development of technically robust and normatively acceptable indexes.

Fourth, values also shape the particular ways an index is constructed, as we have seen in the choice of a dimensional framework to highlight that factors other than the economic are critical to moving toward sustainability. Normative concerns can also directly shape the indicators that are developed or selected for use in an index. For example, some indexes include measures of national energy or financial independence in their sustainability indexes because such independence is culturally valued even though complete independence may reduce supply diversity and therefore sustainability. Normative considerations also play a role whenever a threshold is determined beyond which an activity or state is deemed unsustainable because norms shape the level of risk a group or individual is willing to face (Checker 2007, Gallopin 1997: 15, 17, 19, Obst 2000: 14, Satterfield et al. 2004). For example, nuclear waste containment facilities are designed with technical considerations for shielding radiation *and* the public's willingness to risk its effects in mind.

Constructing an index out of indicators also relies on normative decisions: how to weight individual indicators relative to each other (Becker 2004: 207). For example should subindexes corresponding to the three dimensions be equally weighted, or should one be given precedence? While arguments can be made for prioritizing any of the dimensional subindexes or for equally weighting the components, all aggregation systems will rely on technical and ethical evaluations of how components contribute to the index goals (Becker 2004: 207).

Finally, norms shape the use of indexes. After all, if an index obviously favors something that a community abhors, the index is unlikely to be used. Alternatively, when people perceive that indexes monitor something valued by the community, as many people think GDP monitors well-being, the index is likely to be popular even if its details do not align well with community values. Indexes developed through participatory processes can foster movement toward sustainability as people value what the index monitors and desire its improvement. Indicators may also influence people's actions if they buy into the index and value out-achieving neighboring communities or nations who use the index. Thus, cultural values and ethical priorities can influence the efficacy of indexes.

As we have seen, normative concerns affect indexes at every stage including the development of sustainability definitions, the decision to monitor progress toward sustainability, decisions about the processes by which indexes are created, and intricate decisions about the way indexes are developed and finally implemented. Yet these connections between norms and indexes are neglected to the point that they are not always recognized in lists of considerations for index development. Certainly, there are few if any generally accepted guidelines for using ethics to guide the development or implementation of indexes, and none which are widely accepted. The absence of such guidelines enables the uneven inclusion of ethics in indexes, possibly ignoring some norms altogether, a situation that will hinder movement toward sustainability which is defined by both technical and normative considerations.

3.5 Incorporating ethics into sustainability indexes

In order to guard against the limitations of failing to adequately consider norms in the development and implementation of sustainability indexes, I propose a multipart plan correlated with the five ways ethics relate to sustainability indexes. In this section, the general features of this method will be articulated. They will be elaborated upon and developed in greater detail in the chapters to come.

First, ethical principles for sustainability should be identified and developed simultaneously with technical issues so they can and do mutually influence each other. Such principles should resonate with the people who will be affected by the index. As will be discussed in Chapter 4, the principles should be broad enough to resonate with multiple ethical traditions but specific enough to demarcate broad bounds of what is ethically acceptable to discriminate among options in collaborative decision-making among people of diverse ethical traditions. While groups seeking sustainability indexes may wish to start from scratch to develop their own norms of sustainability, it is likely that an existing list would help many understand ethical reflection while preventing them from reinventing the ethical wheel. Thus, I recommend starting with the ethical principles outlined in Chapter 4, modifying them or adding to the list as needed to capture the important aims of the community's normative priorities. This list is a good place to start because of the broad consensus from which it springs and since it has already been refined in conversation with multiple ethical traditions. Whether or not the ethical principles identified in Chapter 4 are used as a starting point for ethical reflection, intentionally reflecting on ethical aspects of sustainability from the beginning of the process of index development will help ensure that they have a chance to pervade the index.

Second, as index development begins, and throughout this process, participants should recognize that their decision to focus on indexes in general and quantitative indexes in particular as they probably will, is itself a normative discussion. Acknowledging this is the first step toward exploring whether other types of indicators (e.g. qualitative indicators) may be best suited to assess progress toward sustainability goals and then developing and employing them. Such considerations, while critical before the index development process begins, should also occur during the process of identifying and aggregating indexes and when index performance is assessed to ensure that the type of index selected continues to align with normative and technical priorities of the communities using and affected by the index.

Similarly, participants should initially and periodically assess whether the types of participants and means of their participation mesh with the values of index users and enable the collection of adequate data to employ the index. Of course, the position taken should be informed by the normative priorities of technical goals outlined in step 1, but periodical revisiting of this point is necessary because environmental and social situations may change, people may develop new or recognize old normative commitments, and participatory processes may require this revision.

With a preliminary set of technical and normative priorities for sustainability, as well as a type of index and a form of selecting index developers in place, the process of indicator identification and their aggregation into indexes can begin. Typically, the influence of norms at this stage of the process, as indicated by guidelines for index development and discussions of needed tradeoffs in index development, is not explicit. This situation, as we shall see in Chapter 5, often leads to indexes that neglect central ethical elements of sustainability discourse. Therefore, careful work is needed in this area. To do this I propose that index developers study the ethical principles, reflecting on what they imply for sustainability in their cultural contexts and how progress toward the index goals could be monitored. Two broad ways to do this exist. The first involves enabling ethical principles to guide the overall scope of indicator selection. For example, the principle of responsibility suggests that people should know something of the negative impacts of their actions and work to reduce such impacts. Therefore, an index generally focused on responsibility could monitor the emissions of certain pollutants or the change in ecosystems because of these pollutants, for example the rising acidification of lakes downwind of coal fired power plants. Such general influence already occurs, but will not necessarily sufficiently incorporate ethics into indexes when used alone. Enter the second method: specific indicators which monitor the adherence to a particular ethical principle. Continuing with the principle of responsibility, one could monitor changes in the degree to which people say they recognize their culpability with respect to a form of environmental degradation in particular or with respect to environmental issues in general. Alternatively, one could monitor the degree to which people take responsibility for their actions by advocating for different environmental legislation; boycotting companies with poor environmental records or patronizing those with good records; or changing their actions at home or work. Such direct indicators may be necessary to monitor progress toward certain normative aspects of sustainability.

Index developers should also assess whether the data collection or aggregation processes violates ethical norms. Similarly, the evaluation of indexes to ascertain if they are operating as expected and producing useful results to enable progress toward sustainability should be guided by ethical principles and broader normative concerns and suggest ways to guide revisions to the indexes and their implementation.

While these processes are only outlined in general here, the book as a whole demonstrates how these steps can be undertaken in ways that integrate normative and technical concerns in index development. Defining ethical principles for sustainability occurs in Chapter 4 as does the commitment to a type of participatory process. A commitment to indexes is presumed throughout this book and explained most explicitly in the introduction. Incorporating ethics into sustainability indexes and evaluating the degree to which they permeate indexes is explored in Chapters 5–8 which examine the ethical limits of existing indexes and suggest methods of overcoming these limits.

Given all of the opportunities for values to influence the process of developing sustainability indexes as described in Section 3.4 and the relative lack of dialogue between sustainability index developers and ethicists as described in Chapter 2, it is not surprising that index developers have focused much of their time on articulating terms, guidelines, and frameworks as well as noting the polarities of 1) comprehensiveness and manageability, 2) technical rigor and accessibility, and 3) ideal and actual data that must be balanced during index development in general and the development of sustainability indexes in particular. This focus has provided significant structure to a field that barely existed twenty years ago and that integrates the work of so many disciplines (economics, development studies, environmental studies, sociology, well-being) and aspects of life (biodiversity; ecosystem services; air, water, and land pollution; quality of life; human health). In this way, sustainability index studies have been quite successful theoretically and insofar as sustainability indexes are now sought by concerned citizens, consumers, policy-makers, and business and used by thousands of individuals, companies, nonprofits, and governments at local, regional, and national levels. This focus on technically defining the field, however, means that index developers have paid relatively little attention to the multiple ways ethics can influence indexes as outlined above. If this explicit neglect of norms had no substantial effect on sustainability indexes or on making progress toward sustainability in general, it would be of little concern to anyone but theorists. However, as Chapter 5 will show, the inattention to ethics on the part of sustainable index developers can lead to indexes which fail to align with the most deep-seated norms of sustainability, inhibiting their emphasis in policy and action and therefore hindering progress toward a technically sound and normatively acceptable sustainability. Thus, a deeper integration of ethics into sustainability indexes is warranted.

Note

1 Chapter 3 sections 3 and 4 were adapted from the author's article, Fredericks, S. E. (2012). Challenges to Measuring Sustainability. In I. Spellerberg, D. Fogel, S. E. Fredericks, L. M. Butler Harrington, M. Pronto & P. Wouters (eds.), *Berkshire Encyclopedia of Sustainability* (Vol. 6: Measurements, Indicators, and Research Methods for Sustainability). Great Barrington, MA: Berkshire Publishing Co. Adapted with permission from Berkshire Publishing Group.

4 Comparative ethics for sustainability

To thoroughly incorporate ethics into indexes one must explicitly identify ethical content and methods. Yet selecting an ethical system is challenging given ethical diversity and decreasing scholarly and lay acceptance of imposing one ethical system on all people. Additionally, sustainability ethics must both be universal enough to shape transnational policies, since many sustainability issues do not stop at national borders, and align with the particular environmental situations and worldviews of local communities so they support and work toward sustainability. All too often, however, sustainability ethics or environmental ethics in general either 1) aim at universality without considering variations among local ethics, yielding ethical systems or recommendations unacceptable to many, or 2) emphasize the importance of community-based ethics to the point that they overlook the fact that dialogue between worldviews is necessary to address global sustainability issues. A pragmatic theory of ethical principles will avoid these extremes. This method, outlined below, enables diverse groups of people to have a common ground for ethics without requiring them to give up their deeply held beliefs and practices. Its ethical analysis utilizes broad ethical principles that resonate with people of many different ethical traditions and can broadly constrain what actions are deemed ethical even as the principles may be specified in different ways in various traditions. These principles are open to revision over time as new people enter the conversation, as environmental or social situations change, or as knowledge or priorities evolve.

Section 4.1 outlines this pragmatic theory of ethics. Section 4.2 identifies a preliminary set of such principles, beginning with Agenda 21's ethics and specifying them according to ethical positions of three religious and philosophical worldviews, that of James A. Nash, Christian ethicist; Othman Abd-ar-Rahman Llewellyn, an environmental planner well versed in Islamic law pertaining to the environment; and deep ecologists including Arne Naess, Richard Sylvan, and David Bennett. Readers more interested in indicators than philosophical theory may want to skip to Sections 4.1.3 and 4.2.3 to read the summaries of the theory of broad principles and the descriptions of the principles. Section 4.3 examines potential critiques of this comparison. Finally, Section 4.4 outlines a method of employing the principles.

Before proceeding with this plan, however, I will briefly respond to likely questions from environmental ethicists. First, they are likely to ask why I do not just use the principles of the Earth Charter, an international document for environmental and sustainability ethics developed through a multiyear participatory process, if I wish to focus on consensus-based principles of environmental ethics. I choose not to use the Earth Charter's ethics for three major reasons. Its general tone of eco-spirituality can be off-putting both to people who are adherents of traditional religions that are not Gaia-based and people who claim to have no religion. The method described in this chapter can address the concerns of both of these groups more thoroughly. Second, while the Charter was developed through an iterative process, it seems that its supporters have largely if not completely closed the Charter to revisions. While I understand that some periods without revisions may be necessary while building support for such a document, the fact that the Charter is not billed as potentially open to further revision is problematic. It means that people who do not agree with its tenets, tone, or methods are likely to be put off by the whole thing rather than decide that they too can contribute and participate. What if new dimensions of sustainability issues or new ethical responses are identified or developed? What of the new generations who have come of age, or been born, since the Charter conversations took place? Might they see things differently? What of the idea that idealized, static modes of thought including ethics seem to have contributed to environmental degradation in the first place as people were not able or willing to revise their thinking and actions quickly enough to anticipate or respond to new environmental issues? Doesn't the lack of a revisionary process set the Earth Charter up for similar limitations? We get ahead of ourselves if we think that current ethical documents might not be improved upon in the future as new ethical knowledge and methods develop, new conservation partners enter the dialogue or as social and environmental situations change. Additionally, considering ethics as a realm of static ideals rather than a body of knowledge and methods that change over time belies the historical record and problematically sets ethics apart from other human endeavors such as the sciences, which are widely recognized to change. Third, intellectually speaking, the Charter and supporting documents do not sufficiently articulate the philosophical mechanism by which consensus around ethical principles can arise among people of very different worldviews. Without such a discussion, I do not know the degree to which the Charter proponents expect that it will be supplemented by people of different backgrounds. My pragmatic theory, articulated in the first section of this chapter, however, addresses philosophical questions about the scope of the broad principles identified in later sections and illustrates the degree to which they can be specified in different ways by different groups while placing broad bounds on ethical actions. While the Charter might align with a similar theory, some elements of the Charter are so specific in their proscriptions as to make full alignment unlikely. Thus, I do not begin with the Earth Charter as the source for sustainability ethics.

Another question about my methods is likely to come from environmental pragmatists. Namely, if I am concerned with the practical problem of ensuring that consensus-based decision-making among people with divergent worldviews influences environmental decision-making and action through the development of sustainability indexes, why articulate ethical principles at all? Why not just focus on practical problem-solving itself, as a variety of pragmatic ethicists do, claiming that differences in ethical claims and metaphysical assumptions will only bog down discussions and detract from the real business of consensus which can be reached on many practical matters without agreement about ultimate ideals? Bryan Norton, for instance, tends to focus on problems and set aside theological and metaphysical differences with the thought that such differences do not necessarily yield different practical solutions and therefore it is unnecessary to engage with them to any significant extent when devising responses to environmental challenges (Norton 1991, Norton 2005). While I agree that people with different theological and metaphysical commitments do not necessarily make different policy decisions, I disagree that removing such commitments from the conversation altogether is productive. All too often, a pragmatic focus on the problems is accompanied by a tone of dismissal for anyone who holds longstanding or newly emerging metaphysical or religious ideas. Such a tone is not conducive to collaborative dialogue since so many people are profoundly shaped by and expect to reference their worldviews when discussing ethics. How are they to participate if their deeply held beliefs and, indeed, their selves are dismissed by those calling for the conversation? To ensure that individuals' and communities' deeply held beliefs are acknowledged and respected and can be held even as a language of collaborative sustainability ethics is developed, I articulate and implement the theory of broad principles described below. The theory behind these principles, the process of identifying them in this chapter, and the process of implementing them in the next chapter all illustrate that people can retain deeply held, contradictory beliefs, say between biocentrism and weak anthropocentrism, or whether or not a creator God exists, while forging broad principles. Explicitly naming ethical principles can also help ensure that the ethical considerations are not completely overshadowed by technical concerns.

Finally, some readers may wonder whether people with very deeply held convictions about the ground or content of their ethics will acquiesce to using the broad principles, speaking at least sometimes in a more general manner abstracted from their terminology and deeply held convictions. Addressing this question is a critical one but is more easily examined once we understand the details of the theory of broad principles and their content. For now I note that the theory of the broad principles, especially their revisionary nature, their ability to be specified in different ways in different traditions, and the fact that they set broad bounds on what decisions are ethical, but can be interpreted differently within that broad bound in local settings according to local norms, will ensure that their process and content should be acceptable to large segments of the population. To understand these claims in more detail we must turn to the ethical theory itself.

4.1 Theory of pragmatic ethical principles

To develop a theory for sustainability ethics that can be relevant to both international policy-making and the particular ethical traditions of various local communities. I draw upon two strands of twentieth-century scholarship: 1) Robert Cummings Neville and Wesley J. Wildman's extensions of Charles Sanders Peirce's semiotic work in their development of the idea of the "vague category" and 2) Walter G. Muelder's concept of the "middle axiom." Each term brings unique resources to the project of articulating a theory of cross-cultural principles of sustainability ethics and identifying such principles. Explications of the vague category contribute philosophical precision and criteria for identifying commonalities across multiple worldviews, a helpful trait given the global arena of environmental ethics. Meanwhile, Muelder's explicit ethical work and its implementation by the World Council of Churches yield a helpful corrective to the vague category's theoretical focus through concrete examples and the emphasis on implementable and effective ethical principles. Combining and expanding these sources, and switching terminology from vague to broad, I define a broad ethical principle as one that is broad enough to be accepted by people of widely different worldviews yet can be specified in different, possibly contradictory ways according to the beliefs, practices, and environmental settings particular to each worldview. Broad ethical principles also mediate between the concrete actions of a worldview's adherents and the fundamental theological and philosophical assumptions of the worldview. Guided by criteria for application, the broad principles are guides for decision-making and action between and within worldviews. Let us look to each of these strands of twentieth-century scholarship in more detail before combining them.

4.1.1 Peirce, Neville, and vagueness

Grounding Neville and Wildman's idea of the vague category is Peirce's work on semiotics, the theory of signs, particularly the philosophical concept of vagueness (Neville 1992: 25). For Peirce, signs come in three types: icons, indexes, and symbols (Peirce 1931c: 299–300). Photographs are excellent examples of iconic signs because they, like all icons, only use their own characteristics to represent their objects. Icons represent objects that may or may not exist (a photograph of my third birthday cake still represents the cake though it was eaten years ago), but indexes require a causal relationship with the object they represent. Thus, my finger is an indexical sign when I use it to point to something, but not when it is loosely held at my side. Symbols, the third type of sign, represent their object by characteristics established in a community to relate the object and the sign's interpreter (Neville 1992: 33).

For example, the Olympic rings are symbols both of the Olympics themselves and the values of peace and competition associated with the games. Only people a part of the community related to the modern Olympics understand the meanings of the flag; it would have no meaning for people who lived before the start of the modern Olympics. Signs are central to Peirce's philosophical system because they are the means by which all thought occurs (Neville 1992: 26, 34, Peirce 1931b: 169–89).

Peirce classifies signs according to two other terms: the general and the vague. In relation to one of their aspects, each sign must be either vague, general, or somewhere in between (Peirce 1931c: 304–5). Both vague and general signs are indeterminate because they do not themselves determine their referent. The interpreter determines what a general sign refers to, while another sign fills this role for vague signs. Thus, "Man is mortal" is a general sign because it can pertain to any person specified by the interpreter while "an almanac predicting a great event this month" is vague because the almanac does not specify what great event is to occur, and another sign must be used to specify this information (Peirce 1931a: 354–57, Peirce 1931c: 299–300). Indeed, vague signs are often linked in extensive chains of specification. For example, biologists classify living entities according a series of nested signs. The kingdom of *Animalia* may be specified by the *Chordate* phylum, creatures with hollow spinal chords, which may be specified by the *Mammalia* class. Several more specifications are needed until one reaches the level of *Homo sapiens sapiens*, our subspecies. While this chain of classifications grows ever more specific, each level is still vague. No level names a particular entity and those entities the classifications represent could have different, even contradictory characteristics. For instance, *Homo sapiens sapiens* can be blind or sighted. To emphasize the fact that a vague sign encompasses many other signs, I will follow Neville's use of the term "vague category" rather than Peirce's "vague sign" though technically both the category and its constituent parts are signs (Neville 2001a, Neville 2001b, Neville 2001c).

Another major difference between vague and general categories is the degree of fixity of the relationship between their specifications. The generalness of a category implies that all of its possible specifications are united by a clearly defined relationship. For example, all triangles in Euclidian geometry (a general category) are figures composed of three straight sides in a plane whose interior angles add up to 180 degrees (Peirce 1931a: 356). The *vagueness* of a category, however, implies that its specifications are related in possibly more complex ways that can only be understood through extensive trial and error. For example, many religious studies scholars have slowly determined that "God" is not a good vague category to describe all religious traditions because of its deep relation to the claims of particular monotheistic traditions while "Ultimate Reality" or "Ultimate Realities" may be more fitting. Similarly, vague categories are not static entities articulated once but are continually open to testing and then modification or replacement (Neville and Wildman 2001c: 198).

Vague and general categories also differ by their treatment of contradictory specifications. The law of the excluded middle (that a thing can only fall into one of two contrasting categories, A or not A) does not apply to the general category itself. Consider a general category such as the idea of the triangle. It is neither isosceles nor not isosceles, neither equilateral nor not equilateral (Peirce 1931a: 356). Rather, the general category of triangle could be specified by any of these types of triangles, even though as a category it is none of them. The law, however, does apply to the specific entities in a general category: a *particular* triangle is either isosceles or not, and regardless of how it is classified in these terms, it is still a triangle. For example, a 45, 45, 90 triangle with sides 1, 1, and $\sqrt{2}$ meters or an equilateral with 3-meter sides are both specifications of the general category of a triangle. Though these specifications are different, they are not *contradictory* because both specifications could in fact be true for different triangles and fit in the general category of a triangle.

On the other hand, the principle of noncontradiction does not apply to the vague category itself nor to its specifications (Hartshorne and Weiss 1931: 5.505). To use Neville and Wildman's example, the vague concept "subjective experience of all reality" can be specified by "life is a blast" or "life is suffering." While commonsense understanding says that life cannot both be "a blast" and "suffering" each of these contradictory ideas "can meaningfully specify the vague category of the ordinary experience of life" (Neville and Wildman 2001c: 198). Thus, the main differences between the vague and general are that general categories encompass specifications identified by the interpreter that can be simultaneously true in all respects at the level of the category. In other words, they are noncontradictory. Vague categories need to be specified by other signs, possibly in contradictory ways, with relations between the specifications to be determined through time by trial and error.

Because Peirce and Neville think all thinking is done by signs, apply the concept of "vague" to a variety of ideas, (Neville 1987: 93, 100, 135, 128–31, 190, Neville 1989: 26, 54, 141, 170, Neville 1992: 67, 106–8, 206, Peirce 1931a: 347–48, Peirce 1931d: 116–18, 122–23, Peirce 1931e: 317–18, 343–45) and have written on ethics, it would not be surprising if either of these scholars had applied the concept of the vague category directly to ethics. One could imagine a discussion of vague guidelines for action such as "love your neighbor" which could be specified depending on definitions of neighbor, type of love, and the situation in which it is used. Yet neither Peirce nor Neville fully make this move. Peirce focuses his ethical discussions on the hypothetical nature of ethical theories, the relationship of ethical theory and practice, and the relation of ethics to logic and his ontological system (Feibleman 1969: 366–87). Neville does posit a connection between ethics and the idea of the vague category as he notes that moral theories that "select certain things as worthy of description" are themselves "somewhat vague" and will need to be specified by other theories closer to the

78 Comparative ethics for sustainability

phenomena (Neville 1987: 92–94). Additionally, he maintains that cultures embody norms in their concepts and actions and that each may specify its normative rituals differently (Neville 1995). However, he only observes the possibility of vague normative categories; he does not try to identify a set of vague principles that may register the norms of multiple cultures and facilitate ethical decision-making among people of different cultures.

I posit that the vague category can be extended to principles of ethics. Vague principles are a way to conceive of the relationship between ethical principles rooted in different worldviews. The principles of each worldview, though distinct, are specifications of overarching vague principles. To flesh out and employ this theory, a method of constructing vague categories across worldviews and applying them in ethical decision-making and action is needed. The work of the Comparative Religious Ideas Project (CRIP), a project led by Neville, as well as the work of Muelder helps to fill in these holes respectively.

The CRIP sheds light on the process of using the vague categories across worldviews traditionally categorized as religious but could be useful to compare any worldviews (Neville 2001a, Neville 2001b, Neville 2001c). It aimed to state "explicitly how religious ideas differ and how they are the same, where they overlap and where they are mutually irrelevant, in what their importance lies, and what connections among them are trivial" with respect to the vague categories of the human condition, religious truth, and Ultimate Realities (Neville and Wildman 2001a: 3). These categories are vague because the individual religious traditions specify the categories in different, yet possibly contradictory ways (Neville 2001c: xxiv). For example, Ultimate Realities can be specified by definitions of the Ultimate in ontological terms, as in Hinduism's idea of "Nārāyaṇa as the creator of the world on which all else depends," Judaism's portrayal of its God as "the creator of the world and goal of human existence," and the "Confucian notion of Heaven." Ultimate Realities can also be specified by anthropological ultimates that define what is ultimate in relation to humanity's needs or desires (Neville and Wildman 2001b: 1). For example, utilitarianism's is the greatest good for the greatest number. Thus, the vague category of Ultimate Realities is specified with other vague categories, anthropological and ontological ultimates, each of which can be specified in highly divergent ways.

CRIP scholars acknowledge that comparing ideas as complex as religious ideas using vague categories is a difficult process easily prone to distortion by the biases of the one doing the comparison so they articulate a theory of careful comparison criteria to evaluate the results of such comparisons. They maintain that entities to be compared must be understood on their own terms and as they influence ideas and actions both inside of and outside of their semiotic systems. Though entities can be compared in all of these ways, rigorous comparisons should also acknowledge that there is a sense in which ideas themselves cannot be compared because of the limits of translation from one worldview to another. Once comparisons are made, they need to be evaluated for rigor and significance. This is usually done through pragmatic test rather than those of formal logic and using relative criteria including "consistency, coherence, applicability to the subject matter, and adequacy" (Neville and Wildman 2001c: 189). If the comparisons are inadequate, they should be amended as needed (Neville and Wildman 2001c: 190–91).

Through the process of comparison between the specifications of the vague categories, the participants in the CRIP, and those who now read about the project, are both able to refine the vague categories and gain a deeper understanding of the important connections and discontinuities between religious traditions (Neville 2001a: xvii–xxii, Neville and Wildman 2001c: 187–88). This method balances the desire for universal concepts with the respect for the particularities of each tradition, exactly what we are looking for in a theory of sustainability ethics. Yet, this comparative method and the idea of vagueness need to be modified to apply to ethical issues which require applicability in practice.

4.1.2 Muelder and middle axioms

To translate the theory of vague categories to the needs of applied ethics the work of Walter Muelder (1907–2004), a Christian social ethicist and theologian concerned with ecumenism and the application of ethical theories to real life decision-making through the concept of a "middle axiom," will be helpful even if we do not share his theological commitments. For Muelder, ethics consists of "ultimate ideal goods," "the concrete programs of ethical commands," and the significant practical and logical gap between the two which can be bridged by his concept of "middle axioms" (Muelder 1966: 10). According to Muelder, ideals consist of moral laws which are universal in the sense that they apply to nearly all situations unless superseded by a more general law (Muelder 1966: 17). For Muelder, moral laws include "The Law of Specification: all persons ought, in any given situation, to develop the value or values specifically relevant to that situation" (Brightman 1933: 171, Muelder 1966: 52); "The Law of Cooperation: All persons ought as far as possible to co-operate with other persons in the production and enjoyment of shared values"; and the Metaphysical Law: "All persons ought to seek to know the source and significance of the harmony and universality of these laws, i.e., of the coherence of the moral order" (Muelder 1966: 60). Muelder holds that moral laws provide structures for ethical decision-making but do not uniquely determine the outcome of any ethical decision since they are ideals removed from the details of actually making decisions (Muelder 1966: 10, Muelder 1983: 286). Critical details which influence human actions such as who or what is involved, who or what will be harmed, who has the power to make a decision or act, and the history of the situation will always be separated from ideals as are the facts that we cannot fully know, desire, or achieve our ideals due to human fallibility (Deats 1986: 285).

Muelder uses "middle axioms," a term coined by J. H. Odham, to bridge the gap between ideals and concrete actions. Middle axioms inhabit broad bounds set by moral laws though laws do not uniquely determine their content (Muelder 1959: 21, Muelder 1966: 10, Preston 1986). Though middle axioms do not determine a single correct act for any circumstance, they do constrain the bounds of acceptable actions much more than moral laws.

Muelder's primary example of a middle axiom is the "responsible society." It was developed in 1948 when the World Council of Churches (WCC) sought an economic "standard ... relevant to the needs of their members" both in communist and capitalist societies (Wogaman 1993: 257). Muelder follows the WCC's definition of the "responsible society" as "one where freedom is the freedom of men [sic] who acknowledge responsibility to justice and public order, and where those who hold political authority or economic power are responsible for its exercise to God and the people whose welfare is affected by it" (Muelder 1959: 19). In a responsible society all people should be treated as ends, not means, by states and economies since these institutions are created to serve people. The

> responsible society also emphasizes freedom, justice, and equality, mandating that people have freedom to control, to criticize and to change their governments; that power be made responsible by law and tradition and be distributed as widely as possible through the whole community; and that economic justice and equality of opportunities be established for all members of society.
>
> (Muelder 1959: 19)

Muelder refined the nature of the "responsible society" in *Foundations of the Responsible Society*. After articulating the general characteristics of a responsible society, including equality, freedom, and justice, he outlined the implications of a responsible society with respect to economics, farming, social welfare, and other spheres of human life. Yet, as befits a middle axiom, Muelder posited general priorities that make up the responsible society rather than definite prescriptions for social policies. For example, in his section on responsible consumption, Muelder favored using "qualitative rather than predominantly quantitative standards" because he thought that quantitative measures usually monitor how much we consume without considering whether this level of consumption is just or equitable (Muelder 1959: 242). Thus we see that Muelder's description of the responsible society is more concrete than moral laws because it identifies ways that the laws may be followed in concrete actions even though it is not specific enough to uniquely determine ethical actions.

Because middle axioms highlight common features of moral systems that may have contradictory specifications, they can be classified as vague categories for ethical guidelines. Admittedly, Muelder did not ground his ideas in semiotics, work out detailed theories of how we develop middle axioms, or discuss how they relate to each other in as much detail as Neville and Wildman do. His use of middle axioms does, however, reveal two key modifications of the theory of the vague categories particularly useful for ethics.

First, Muelder's work emphasizes a more intricate picture of the relationship of levels of middle axioms than is typically discussed with respect to the vague categories. As you may recall, each vague category can be specified by multiple categories, each of which can have multiple categories specifying it. At each level, the categories may have contradictory characteristics and yet equally specify the category vaguer than it. To specify one of these categories, we must choose *one* specification at each level, as an animal belongs to one kingdom, one phylum, one class, one order, one genus, and one species. Similarly, Muelder posits that "there may be several levels of generalization or abstraction in this middle range of moral propositions [the middle axioms]," though his examples reveal a more complex relationship than a simple hierarchy (Muelder 1966: 10). For instance, Muelder names justice, equality, and freedom as the components of the responsible society (Muelder 1966: 33). These three components, possibly middle axioms themselves, are mutually dependent on each other and together are needed to specify the responsible society. Thus, he maintains that two people implementing the middle axiom of the responsible society in different cultural contexts will each do so through justice, equality, and freedom though their definitions of these terms may vary. The way they define each term will shape their definitions of the others and of their vision of the responsible society as a whole.

Recognizing the complex interrelationship of vague principles of ethics and their specifications highlights two important characteristics of using vague principles in ethics. First, it provides a framework for discussing the points of consensus and disagreement between worldviews. Second, it enables us to acknowledge that ethical principles at any one level of abstraction may be intertwined, as often occurs. Thus, ethical categories, axioms, or principles ought not to be applied in isolation if one aims to apply them in ways consistent with actual practice.

Muelder's focus on applied ethical decision-making also necessitates adding the criteria of tractability to the criteria of comparison (consistency, coherence, applicability to the subject matter, and adequacy) articulated by Neville and Wildman (Neville and Wildman 2001c: 189). They are focused on faithfully representing the traditions under study. Muelder's goal, of course, is to develop and encourage the use of the responsible society and other middle axioms to establish ethical norms appropriate in multiple traditions to guide decision-making. Indeed, a significant portion of *Foundations of a Responsible Society* is devoted to demonstrating how the middle axiom of the responsible society can shape decision-making. Here Muelder follows a criterion of tractability which prioritizes middle axioms that can helpfully guide decision-making – those that are related to the world in which decisions need to be made, resonate with moral traditions, and productively constrain actions. Similarly, as we seek ethical principles that can guide sustainability discussions and actions we must strive for tractability in any vague ethical category we articulate. Combining Muelder's insights with the well-articulated concept of the vague category yields a theory of broad ethical principles that can promote ethical dialogue across worldviews.

4.1.3 Broad principles for ethics: combining insights of vague categories and middle axioms

I define a "broad principle" for ethics as a guideline that mediates between ideals and concrete reality and is broad enough that it can be specified in different, possibly contradictory ways depending on the details of the situation in which it is used. Such broad principles can only be identified through a careful process of comparison in which people within and outside of the traditions test the proposed broad category against the traditions it supposedly registers and for tractability. After broad principles are articulated, they can be applied as an interconnected set to guide consensus-based moral action across multiple worldviews. My choice of the term "broad principle" to unite the rich heritage of the middle axiom and vague category is quite deliberate.

I choose "broad" instead of "vague" to ensure that readers, especially those not familiar with this philosophical tradition, do not focus so much on the pejorative connotations of "vague" that they cannot appreciate its utility when describing a category of ethical principles. The ethical principles are not so vague that they inhibit meaningful ethical analysis. Examples of their application in Chapter 5 will help overcome such presuppositions, but using a term other than "vague" may help a reader get to these examples. I do note, however, that the negative implications of vague is *appropriate* in a philosophical context if applied to a group of ethical principles that are not similar in a meaningful way (Neville 1992: 146, 198–99); this caution can be incorporated into the definition of a "broad principle" without using the term "vague" which is likely to be dismissed out of hand by a mixed audience of ethicists, index developers, and stakeholders.

With the choice of the term "broad" I also deliberately distance myself from those who, drawing on the terminology of "thick" and "thin" descriptions in anthropology use thick and thin to describe ethics, where "thin" is used in parallel with broad and thick is used for the specifications. I've never been fond of these terms for ethical categories as discussed here because "thick" is not as able to connote an overarching, encompassing, umbrella category that links various specifications with common themes as "broad" can.

I choose the term "broad principle" rather than "broad axiom" because "principle" avoids the universal, self-evident connotation that is often implied by and assumed of "axiom" (though Muelder does not use it in this way) (Bennett 1946: 77). After all, it is only after careful consideration that a principle can be considered broad with respect to specific characteristics of multiple worldviews. To even more assuredly avoid the static certainty of

Comparative ethics for sustainability 83

"axiom" I could use "guideline" instead of principle. Guidelines are considered as more suggestions than rules; they are often understood to change over time as new information is obtained or new situations develop. I do not use the term "guidelines," however, because the principles I have in mind are not as easily set aside as guidelines. While they may be modified and some may even be jettisoned, this only occurs after careful comparison within particular traditions or in collaborative decision-making that leads to the broad principles.

"Principles" are advantageous for cross-cultural environmental conversations because they are often articulated in such a way that they reference the particular values and metaphysical assumptions of a worldview but do so in language that is more readily communicable to outsiders than the ethical stories, tales of moral heroes and assumptions themselves. Thus, focusing on principles, however formulated in individual traditions, can be a first significant step toward rich comparisons and the formulation of broad ethical principles. In particular, broad principles establish a moral language useful to people of different moral worldviews. If used, such a language will not only foster communication among diverse worldviews, but will also enable morally acceptable decision-making about environmental issues in ethically heterogeneous groups as is necessary to address regional and global dimensions of environmental problems.

Despite these advantages, the use of principles can risk imposing the ideals of the powerful on others, losing the particularities of the tradition from which they arose, and reducing rich ethical principles to bland and powerless common denominators. Several strategies including a careful process of comparison by a diverse group of participants who leave the principles open to revision over time help avoid these dangers.

Neville and Wildman's guidelines for establishing categories of comparison, with a few additions, are useful to guide the process of comparison. They assert that each entity to be compared must be understood on its own terms, and as it influences other ideas and actions, both inside of and outside of its semiotic systems (a group of interrelated signs) (Neville and Wildman 2001c: 202–3). Such criteria can help ensure that the broad principles do register important elements of the various worldviews, a necessary step if the broad principles are to be technically sound and ethically resonate with people of multiple ethical systems. Additionally, as we saw in Muelder's work on ethics, each ethical principle must be tractable for environmental decision-making; each one must help people discriminate between options during decision-making and action, either on its own or in combination with other principles. Ethical principles must also be capable of making recommendations that are implementable in particular cultures and environments. For example, a broad principle regarding resource use would need to resonate with both 1) the idea of "eco-kosher" proposed by Arthur Waskow in which Jews connect all consumption, not just eating, with the holy, and which consists of "constantly moving standard[s]" challenging people to damage the earth less than they did previously (Waskow 2003: 313), and 2) secular initiatives promoting energy conservation such as tax rebates for purchasing energy-efficient appliances, the marketing for which usually focuses on the amount of money consumers will save by reducing energy use.

Additionally, the process of identifying principles should be a collaboration among people who are experts in different worldviews to ensure that their distinct features are well represented. If that cannot occur, as in this single-authored project, care should be taken to examine multiple positions that are different enough to bring challenges to the comparison yet to which the author can do justice. While a single-authored comparison has its limits it does enable the comparative process to be modeled and, of course, its results are open to revision over time. Openness to revision is important if one person or many people are trying to articulate a preliminary set of broad principles. Changes may be necessary as new worldview partners enter the comparison, as new scientific or normative knowledge is gained, and as social conditions and broader ecosystems change. Reviseability also guards against the problems of imposing ethics on others. Indeed, if the principles are understood to reside between concrete decisions and ideals then it makes no sense to put them on a pedestal or force them upon others. At most they can be recommended.

This process enables the identification of broad principles that mediate between ideals and actions and register important elements of multiple ethical systems while preserving their unique interpretations. The broad principles allow individuals to collaborate without becoming hopelessly mired in disagreements over which foundation is best or correct, just what is needed for sustainability ethics.

4.2 Ethical systems

With a theory of broad principles it is now possible to identify a preliminary set of broad principles, the next step toward incorporating ethics into sustainability indexes. Since our goal is a set of principles which reflect normative priorities of the sustainability movement that can be specified by participants from multiple worldviews so that widespread consensus about general normative bounds for sustainability and sustainability indexes can be reached while enabling people to adhere to their longstanding and deeply held worldviews, drawing upon ethical systems widely used in practice will yield principles most conducive to collaborative decision-making. Thus, I begin this comparison with the ethical principles embedded in Agenda 21 (farsightedness, adequate assessment of the situation, adaptability, cooperation, efficiency, responsibility, and equity). After all, it was constructed through a broadly participatory process and, unlike the Rio Declaration and Earth Charter, has been quite influential on index development. Yet, I do not stop with the principles in Agenda 21 because they are not necessarily developed enough to be implementable and because they are not necessarily broad enough to encompass multiple normative perspectives for sustainability. Comparing the principles of Agenda 21 with that of multiple traditions will yield a set of broad principles. In theory, such a comparison should include all operating worldviews and ethical systems. In practice, however, such an extensive comparison would take up multiple volumes, and put off the study of how ethics can be interwoven with technical concerns in the evaluation and construction of sustainability indexes, the main goal of this project. Thus, I selected several ethical systems for comparison which differ according to ethical method and worldview, recognizing that many others could have been involved in this comparison and should be in the future.

The ethical systems chosen all explicitly articulate ethical principles and balance attention to the environmental situation with a commitment to philosophical and theological positions but come from different worldviews, both secular and from "traditional" religions, and use different methods of reasoning, authorities, and presuppositions. These systems include that of James A. Nash, a Christian environmental ethicist who developed ecological virtues from a study of Christian love using a modified natural law theory that takes reason, embodiment, and the natural and social sciences seriously; the work of Othman Abd-ar-Rahman Llewellyn, a scholar of Islamic law who draws upon its traditional sources and methods as well as specific examples of positive law to characterize methods and guidelines of Islamic environmental law; and the work of deep ecologists including Arne Naess, and Richard Sylvan and David Bennett, philosophers who contributed to the development of deep ecology, an environmental worldview focused on the intrinsic value of all biota and intuitions or feelings of the environment, often coupled with rationality, which has been influential for and resonant in a number of environmental groups including Earth First! (Kamieniecki et al. 1995: 315, Sylvan and Bennett 1994: 145, Taylor 1995: 15–16).

To ground the comparison of these systems to identify the preliminary broad principles, Section 4.2.1 provides a brief background to the ethical systems of Nash, Llewellyn, and deep ecology. With this foundation laid, Section 4.2.2 undertakes a rigorous comparison, broadening Agenda 21's responsibility, farsightedness, adequate assessment of the data, and adaptability; replacing Agenda 21's equity with justice and efficiency with careful use; subsuming cooperation under adequate assessment of the situation and identifying a new principle, feasible idealism. Section 4.2.3 summarizes the results of this comparison in a working list of the principles.

4.2.1 Introduction to the ethical systems

Since a rigorous comparison to identify broad principles requires attention to the similarities between worldviews and the ways in which they are distinct, a brief introduction to the methods and content of the ethics of Nash, Llewellyn, and deep ecologists will set the stage for the comparison in later sections.

While one might expect that a Christian environmental ethicist would look to the Bible, and more broadly look to the Christian tradition as ethical sources, Nash maintains that the Christian tradition, particularly the Bible, has few unambiguously positive portrayals of or positions toward nature. Thus, while he relies on the Bible to some degree, and traditional theological concepts including sin, love, salvation, and incarnation to a greater extent, he also utilizes natural law theory and scientific data to reformulate Christian ethics in light of contemporary ecological problems (Nash 1991, Nash 2000: 227–28). Traditional natural law theory utilizes reason to identify ethics; Nash modifies the natural law tradition to reflect human and biotic conditions and draw upon insights of natural and social sciences rather than focusing on "pure" rationality as traditional natural law tends to do. In this way, he hopes to partially transcend "arbitrary preferences" (Nash 2000: 231, 233–35) and develop virtues that could be identified by and meaningful for all people, though he focuses on Christians. For example, Nash relies on knowledge of many physical and natural sciences as he explores the ethical challenges of pollution, global warming, ozone depletion, population explosions, species extinctions, and genetic engineering. He also looks to the social sciences for the causes and effects of human population explosions and increasing consumption, as well as the intricate connections between economics, ecology, and politics (Nash 1989: 32–33, Nash 1991: 23–67, Nash 1992: 774, Nash 1994: 140–44, Nash 1995, Nash 1996, Nash 2000: 225, 243–46, Nash 2001). Drawing from these sources, he articulates nine ecological virtues (sustainability, adaptability, relationality, equity, frugality, solidarity, biodiversity/bioresponsibility, sufficiency, and humility) that form the basis of his environmental ethic (Nash 1991: 63–67, Nash 1996: 9).

Othman Abd-ar-Rahman Llewellyn, an environmental planner well versed in Islamic law pertaining to the environment, also aims to expand his tradition's ethical schema to better respond to environmental crises. Llewellyn, however, thinks that Islam has a number of resources for environmentalism. He draws upon *Qur'an* and *ḥadīth*, stories and sayings of the Prophet which are authoritative in Islamic law. He also relies upon the revitalization of legal methods including *ijtihād*, an ancient practice of reasoning from legal precedence and the case at hand to arrive at answers to novel problems, to extend traditional Islamic norms to contemporary environmental issues (Llewellyn 1984: 29, Llewellyn 2003: 193). Llewellyn advocates a return to the "ultimate purposes or objectives of the *Shari'a*" to construct environmental norms (Llewellyn 1984). With this method he aims to avoid both the problems of traditionalists who strictly apply ancient principles about conservation to contemporary situations even though today's problems differ in scale, and possibly, in kind, from ancient concerns, and the problems of reformists who tend to pick and choose environmentally friendly laws from any of the traditional schools of law without articulating a coherent approach (Hallaq 1995a: 207–54). Llewellyn strives to split the difference by following a long history of Muslim jurists who claim that "the fundamental purpose of the

Shari'a is the welfare (*maṣlaḥa*) of Allah's creatures" (Llewellyn 1984: 29, Llewellyn 1992: 89). To discern how to live out the *Shari'a* he relies on traditional Islamic legal instruments (e.g. *waqf*, *ḥimā*, and *ḥarām*), methods (e.g. *ijtihād*), and specific laws as well as contemporary knowledge about the state of the environment (Llewellyn 1984: 29, Llewellyn 2003: 193). Developing Islamic environmental law does face significant challenges including the need to reestablish the use of Islamic law; determining its relationship to secular legal systems and knowledge; constructing the laws; and educating people about and enforcing such laws (Coulson 2003: 47, Haq 2003: 128, Llewellyn 2003: 236, Nasr 1993, Nasr 1996, Nasr 2000). Despite these challenges, many Islamic scholars agree that Muslims must focus on environmental law to have an appropriate Muslim response to growing environmental concerns since Islamic law is the seat of morality in Islam (Haq 2003: 142–50, Llewellyn 1984, Llewellyn 2003).

Contrasting with Nash and Llewellyn who develop new responses to environmental issues while firmly rooted in particular, long-established religious traditions, is deep ecology, a worldview articulated by Arne Naess, a Norwegian philosopher, and has been clarified and extended by other philosophers including Richard Sylvan and David Bennett. While there are many theories of environmental ethics detached from particular longstanding religious traditions, I chose deep ecology for comparison in this study because it has been influential on environmental movements to a larger degree than many of the more abstract theories. Ethical theories that actually matter to people must be discussed when developing sustainability policies and indexes so indexes align with sustainability goals and are accepted and subsequently supported by everyday people. Deep ecology also provides diversity for the comparison as it is a newly emerging position and because its focus on intuiting how to live in relation to the world differs considerably from Nash's and Llewellyn's methods.

Deep ecologists focus on the intrinsic value and interdependence of all entities and how human intuition or feelings of these characteristics of nature shape both human actions and selves. Sylvan and Bennett follow the basic platform of deep ecology but aim to clarify it by rejecting several of Naess' ideas to increase their system's logical consistency and applicability. Specifically, they maintain that certain morally relevant traits correspond to a higher level of moral concern; entities with the same morally relevant traits are to be treated similarly regardless of species (Sylvan and Bennett 1994: 137–41) rather than Naess' idea of biospheric egalitarianism.

Deep ecologists take a variety of positions with respect to modern science. They are often wary of it because of its role in environmental destruction through the application of technology, but recognize that it can inform their assessment of the situation. Deep ecologists also rely on their intuition (Naess) or feeling (Sylvan and Bennett) of the valuing of nature, which in Sylvan and Bennett's case is developed through rationalistic analysis to yield their conclusions. This direct experience of the value of nature is quite influential for deep ecology, and is certainly much more prevalent in deep ecology than in Nash's and Llewellyn's work.

Thus, these three ethical systems vary with respect to their links to traditional worldviews, from reapplying traditional themes and methods to new circumstances (Llewellyn) to creating new virtues and modifying old ones as inspired by the combination of ancient theology and contemporary theology and science (Nash), to developing a new worldview and ethical guidelines (Sylvan and Bennett). They also differ with respect to their sources for ethics as they use some combination of reasoning, Christian and Muslim theological traditions, Islamic law, contemporary science, and intuitions. Additionally, the content of their ethical claims, as discussed below, focuses on different aspects of environmental and sustainability issues. Nash places emphasis on Christian love and knowledge from the contemporary sciences; Llewellyn on revitalizing the methods of traditional Islamic law to fit contemporary environmental contexts; and the deep ecologists on identifying intrinsic value through intuition or reasoning. All of these differences make the three systems a good set to compare when seeking to develop broad principles of ethics as they collectively provoke revisions of the preliminary broad principles from Agenda 21 and represent significantly diverse positions. Before undertaking this comparison, however, we should know something more about the content of each system of ethics.

Nash relies on two primary sources as he develops his nine virtues for environmental ethics: Christian theology centered on the concept of love, and a study of actions necessary to promote ecological integrity in today's world drawing upon natural law ethics, scientific knowledge, and embodied experience. He names love the unifying and motivating theme of all Christian theology, ethical reflection, and action. Love is central for Christianity, asserts Nash, because Christians believe that God is love, a claim emphasized by the prominence of love in the gospels and that "the process of creation is itself an act of love. All creatures, human and otherkind and their habitats are not only gifts of love but also products of love and recipients of ongoing love. Everything then has value imparted by the Source of Value" (Nash 1991: 140–41). Adding support to this idea is the belief that "the story of God's love provides the 'basic moral standard', the 'pattern and prototype', for Christian ethics" (Nash 1991: 141). Thus, while Nash relies on many Christian theological concepts to develop his environmental ethics, he classifies all of these doctrines as expressions of love. For Nash, Christian love

> is always at least caring and careful service, self-giving and other-regarding outreach, in response to the needs of others (humans and otherkind), out of respect for their God-endowed intrinsic value and in loyal response to the God who is love and who loves all. It seeks the other's good or well-being and, therefore, is always other-regarding (only the degree is up for debate).
>
> (Nash 1991: 145)

Thus Nash thinks that love entails recognizing the value of all others on their own terms, because God created them (Nash 1991: 107, 153–54), as well as doing good and avoiding harm, "on behalf of the well-being of others, human and otherkind, simply because a need exists" (Nash 1991: 153). In order to love appropriately Nash claims that humans must know of the needs of our loved ones, both "cognitive[ly] and emotional[ly]" (Nash 1991: 157) and through humility, recognize what knowledge and character one has, may have and can not possibly have (Nash 1991: 63, 66, 156–57). Human limits should also be recognized according to Nash's vision of love so people recognize that the world apart from and including humanity is necessary for our "physical existence, but also for our spiritual well-being" (Nash 1991: 155). Beyond these dimensions of love is communion, the desire for others to be "our loved ones in fully reconciled relationships" (Nash 1991: 157). Such communion, is, however, not complete in history according to Nash; it will only be fulfilled in "The Reign of God" which is the "consummation of communion or reconciliation" (Nash 1991: 160).

While all of the dimensions of love named above are important, the final dimension, justice, is the most important for Nash. Because justice is central to Jesus' message and the Bible as a whole, Nash argues that it is a moral imperative for all who find the Bible normative. It is a minimal expression of love; fully loving inspires much more than justice, but necessitates at least justice. He examines biblical portrayals of God as the "lover of justice"; mandates for justice in both testaments, including the special considerations given to widows and orphans as a part of the relationships promised in biblical covenants; and Jesus' vision of the Reign of God (Nash 1991: 163–65) concluding that justice must focus on the weakest whose rights are easily abused. He sees no reason why justice cannot be extended to all creatures, particularly since the Noachian covenant established that otherkind are included in God's concept of right relationship (Nash 1991: 165–66). Indeed, Nash claims that identifying the proper application of justice to all life is one of the primary tasks of an environmental ethic (Nash 1991: 166–68).

Nash focuses on distributive justice, "the proper apportionment, or allocation of relational benefits and burdens," because it is applicable to all entities, unlike communicative justice (Nash 1991: 164–66). Justice, for Nash, can only be determined within the parameters of a particular situation because justice involves making sure that each entity is given its due, a process of identifying and balancing rights of all involved parties, something that cannot be determined in the abstract. Despite emphasizing the importance of *human* environmental rights, Nash maintains that such focus is insignificant because it does not adequately recognize the value of all entities and will not work quickly or effectively enough to solve environmental crises, especially if economic and environmental considerations conflict (Nash 1991: 172). Thus, Nash extends justice and eight basic rights to all entities with "conation – a striving to be and to do" because "beings may be said to have 'interests' in their biological roles *for their own sakes*" (Nash 1991: 178, 154–55, 186–88,

90 Comparative ethics for sustainability

Nash 1993: 147–48). He is careful to point out that humans have many rights above and beyond this most basic list, that there is no logical reason (though there may be a psychological reason) why biotic rights should diminish human rights, and that entities should be treated the same only when they have similar morally relevant characteristics. Thus he does not worry about "voting rights for chimpanzees – let alone fair housing rights for parasites in human bodies" (Nash 1991: 174). By naming biotic rights and their limits, Nash articulates the minimal bounds of just living – responsibility for ensuring that the rights of humans and otherkind are protected – and preserving necessary conditions for living a life of love (Nash 1991: 167–68).

While Nash's discussions of justice can get fairly specific, he recognizes that most of his analysis of the dimensions of love are still so abstract that they do not readily aid decision-making and action in the complex real world. In Muelder's terms, he needs some middle axioms to link the ideals of love to everyday life. To do this, Nash articulates nine ecological virtues (e.g. adaptability, equity, frugality, humility) that resonate with the dimensions of Christian love and can guide action (Nash 1991: 197–221). Each of these virtues is a mean between extremes which, when practiced over time can shape and become a part of one's character.

Though Nash, like many in Euro-American worldviews and legal systems, separates ethics from the law, Islamic traditions integrate ethical principles, religious rules, and laws about matters as diverse as property, international affairs, and marriage (Llewellyn 2003: 186–87, Schacht 1982: 1). Indeed, if one wants a guide of how to live in Islam, one must look to Islamic law, rather than some discipline called ethics. Thus, it is not surprising that Llewellyn's work to develop an Islamic position on acting in and toward the environment differs methodologically from Nash's environmental ethic though they share many priorities, emphasize the use of modern science, and yield several similar environmental ideas. To understand Llewellyn's position, let us briefly examine foundational beliefs of Islam, theories of Islamic law, and particular methods, principles, and laws deemed helpful for environmental ethics.

At the foundation of the Islamic faith is the concept of *tawhīd*, or unity. As an attribute of God, *tawhīd* connotes the oneness of God. For Muslims, God is the sole and ultimate authority and owner of all of creation; God has created all creatures in a perfect balance so that all may be sustained through their relationships with each other (Kettani 1984: 67). Through these relationships every being fulfills its God-given role to sustain others and worship and serve God (Llewellyn 1992: 89). Consequently, Llewellyn claims that "All beings are ... united in aim, and benefiting the whole is a value that pervades the universe" (Llewellyn 1984: 29–30). Indeed, some claim that the central aim of Islamic law is acting to ensure "the universal common good of all created beings, both in this life and the life after death" in order "to preserve the welfare of Allah's creatures" (Llewellyn 1984: 29, Llewellyn 1992: 89).

Although all creatures are believed to be united by God's order, Islam holds that humans have a special role in the world assigned by God: to be *khilāfa*, a vice-regent or steward. Each human is a *khilāfa* because of his or her "enormous ability to do both good and evil; with ability comes responsibility" (Llewellyn 2003: 190). As a *khilāfa*, Llewellyn argues that a Muslim must uphold "the maxim 'There shall be no injury and no mutual infliction of injury' which protects a person from injury and prohibits him from causing injury to his neighbor, to society or to the creation as a whole," "one of the most important universal principles found in *ḥadīth*," (Llewellyn 1984: 33). Thus, humans are to maintain and care for creation, and, where possible, improve it (Chishti 2003: 75, Clarke 2003: 97).

Through their role as *khilāfa*, Muslims are to embody the ideals of "do no injury" and prioritize actions that will bring the greatest overall good to creatures. To aid the decision-making process, Islamic jurists have developed several other ideals: the good of an action should outweigh the harm that results from it, universal needs trump individual needs, more significant needs should be prioritized over lesser needs, and people with less power deserve special consideration. In Islamic law, these ideals have been most fully developed in the realm of social justice. For instance, an ancient law to promote social justice mandates that the poor be able to collect grain missed by the harvesters and prohibits harvesting methods that prevent the needy from getting their share of the crops (Llewellyn 1984: 33). Certainly, this law would ensure that a greater number of people had access to food, a significant requirement of social justice.

Islamic environmental law often works for social justice to maintain continuity with traditional Islamic law. Indeed, Islamic environmental scholars have already begun to study issues which link social and environmental justice including population control, women's rights, vegetarianism, and prohibitions against waste and pollution (Ammar 1995, Foltz 2003, Hamed 2003, Majeed 2003). In these studies, traditional laws, such as those about gleaning, would need to be revised for application today if the intent of ancient laws will be maintained given the recent changes in social structures, industrialization, knowledge and environmental conditions. For instance, the needy may be far distant from grain production centers.

To understand how these new issues are raised in an Islamic context, we must understand something more about the *Shari'a*, Islamic sacred law, which not only denotes the laws themselves but also connotes "the Way, the path to water, the source of life" (Llewellyn 2003: 187). Muslims believe that by following the law they "live life ... in the most moral and ethical way" according to God's will (Llewellyn 2003: 187). For Muslims, acting ethically is not a trivial matter; their actions may have serious positive or negative consequences both in this world and in the next.

Muslims believe that the ultimate reason to obey the law is God's will. All authority to legislate in Islamic law comes from God. God's will is most powerfully revealed in the *Qur'an* and *ḥadīth*, collections of sayings and actions of the Prophet (and Imams for *Shi'is*). However, Muslims generally believe that the *Qur'an* and *ḥadīth* need to be interpreted to determine exactly what is to be done in every situation. Islamic law developed as Muslims asked *mujitahid*, jurists, how to act in particular situations. They elaborated upon the *Qur'an* and *Sunna* by referring to *ijmā*, the consensus of other jurists, and using *ijtihād*, the hard work of reasoning to determine the law. *Ijtihād* included reasoning to a solution for a novel case from analogies with old ones (*qiyās*), and "judging according to the hierarchy of Islamic values by preferring the stronger or universal values over the weaker or instrumental values (*istihsan*) and where there is no precedent, making judgments on the basis of public welfare (*al-musahh*)." *Shi'is* replaced analogy with reason, *'aql*. (Hallaq 1995a: 1, Llewellyn 1984: 37, Schacht 1982: 1, Weiss 1998: 112, 114, 122).

As Islamic jurists worked to determine God's will in and for the law by relying on the four sources (*Qur'an*, *Sunna*, *ijmā*, and *ijtihād*), various factors led to multiple interpretations of the law. Individual jurists emphasized different portions of the sources of law and used their own reasoning, which certainly differed from person to person. Conventions of local cultures, including class-consciousness, views about women, and the willingness to learn from foreign legal systems also contributed to legal divergence (Coulson 2003: 48–50). From the 1300s, four schools dominated Islamic law: Hanbali, Maliki, Shafi'i, and Hanafi. While all agreed that the *Qur'an* and the *hadīth* were the most authoritative sources of law, they emphasized different *hadīth* and components of *ijtihād* (Coulson 2003: 71–73). Sometimes these differences led to contrasting laws about a particular subject such as developing "virgin" land. Hanafi jurists required land developers to have permission from the local authorities; Maliki jurists required such permission only when the development may harm public welfare; Shafi'i jurists did not require permission at all (Haq 2003: 200–1). Additionally, sometimes one school focused more than others on a particular topic, as when the Maliki school developed the most detailed laws about *himā*, traditional areas of land preservation (Haq 2003: 128, 144).

Though significant variations in the theory and practice of Islamic law exist, several factors avoided fragmentation of Islamic communities along legal lines. First, jurists worked from the same general sources and agreed upon major tenets of the faith such as the five pillars of Islam. Second, when Islamic law was regularly and fully practiced, the majority of Muslims believed that the diversity of opinions in Islamic laws should not merely be tolerated but rather embraced as a sign of divine blessing (Weiss 1998: 116). Third, the dominant theory of consensus helped connect the Islamic community. Bernard G. Weiss describes this majority view as the belief that God intended a particular correct interpretation of the law but that fallible humans were likely to disagree on its content. Supporting this belief is the Muslim adage that God will give jurists a double reward if they are correct and a single reward if they are wrong. Where differences of opinion existed despite a full examination of the sources, the majority believed it was impossible to determine "which opinion, if any, was correct" (Weiss 1998: 119).

Thus, while jurists regarded their authority as deriving from God's, they also usually acknowledged the distinction between their interpretations and God's intended law (Weiss 1998: 120). This division between the ideal law and the law as discernable to humans enabled Muslims to tolerate significant amounts of legal variation. To a contemporary, nonMuslim ear, it may sound as if Islamic law could be stretched to an extreme form of relativism where any interpretation could be acceptable. Yet the shared adherence to sources of law, methods of interpretation, views of consensus, as well as the practical structure of the schools restricted possible interpretations of Islamic law and addressed questions of divergent legal opinions.

In addition to the permissible and even encouraged diversity within Islamic law, the law, especially for those schools that emphasized *ijtihād*, was also quite dynamic in its ability to adapt to new situations while remaining faithful both to traditional beliefs and specific positive laws. Though Islamic legal scholars in approximately 900 CE declared that the "gates of *ijtihād* were closed" implying that all necessary legal decisions had already been determined, recent scholarship suggests that *ijtihād* has always been a part of Islamic law even if less prevalent than before (Hallaq 1995b: 3). This flexibility along with specific ancient laws about social justice, charitable trusts, waste, land use, and water rights could enable Islamic law to directly address contemporary environmental challenges by extending ancient laws to analogous contemporary situations (Haq 2003: 144, Llewellyn 2003: 208, 210).

Though centuries old, widely recognized worldviews such as Christianity and Islam can certainly influence present-day ethical actions and policy-making expectations for sustainability, so too can relatively new worldviews with a relatively small number of adherents. While there are many such worldviews developed in theory or practice, here I look at the deep ecology movement which has both an academic component and a following in the wider world.

In the early 1970s Arne Naess, a Norwegian philosopher, introduced the terms "shallow" and "deep ecology," to denote ways in which people relate to the environment based on their intuitions, ultimate beliefs, and ecological knowledge (not the science of ecology) (Naess 1973, Sessions 1995: xii, Sylvan and Bennett 1994). According to Naess, the shallow movement focuses on "the health and affluence of people in the developed countries" by resisting "pollution and resource depletion" while deep ecology is characterized by biospheric egalitarianism; self-realization; holism; and the prioritization of diversity, symbiosis, complexity, decentralization and anticlassism (Berry 1995: 15, Naess 1995d: 151–52). Later, Naess, and other deep ecologists divided the platform of deep ecology, shared by many with different metaphysical commitments, from those particular to Naess (Naess 1989, Naess 1995b: 214).

The most well-known feature of deep ecology is its emphasis on intrinsic value, the idea that every entity has value in and of itself, not just because humans or other animals use them for food, shelter, or ecosystem services such as cleaning the air or water; because humans think they are beautiful to look at; because humans might want to use them later; because humans like to know that they are there. Aside from its focus on intrinsic value, the core platform shared by most deep ecologists claims that "Humans have no right to reduce this richness and diversity except to satisfy vital needs" and that "The flourishing of non-human life *requires* a smaller human population." The platform also asserts that since the situation is bad and getting worse people who share its tenets are obligated to change policies and social ideologies (Naess 1995a: 68).

Naess himself grounded this platform on a metaphysic in which the interconnection between entities is described as the "relational, total-field image" in which entities are "knots in the biospherical net or [a] field of intrinsic relations" as opposed to distinct entities (Naess 1995d: 151, Sylvan and Bennett 1994: 153–54). For Naess, it is through the process of self-realization that we become one with the world and more fully ourselves (Naess 1995c: 226). Empathy arising from this process combined with the knowledge that our very selves are constituted by relationships with people, places, and all of nature suggested to Naess that we must act on behalf of the environment (Naess 1995c: 226–27, 231).

Naess' ideas about preferred human action regarding the environment tend toward ideals rather than implementable principles. For example, all environmental actions, according to Naess, must be guided by the ideal of "biospheric egalitarianism," treating every entity equally (Naess 1995d: 151). Naess qualifies this ideal with the phrase "in principle" as he realizes that living creatures must eat and thereby kill some and not others – radically unequal treatment at the individual level. Yet after this qualification, Naess does not describe how one can be egalitarian or even move toward this ideal when one must eat, breathe, and do all sorts of other actions which may harm others. Similarly, when Naess prioritizes diversity, symbiosis, and complexity in the environment because ecologists recognize their importance in thriving ecosystems, he does not address how humans are to achieve these goals given that biologists have a difficult time determining what levels of complexity, diversity, and symbiosis are preferred for an ecosystem. Overlooking these concrete issues makes it difficult to understand how to fully apply Naess' theories, unless like Naess, one thinks that if we become whole people we will intuit how to live appropriately in and with the world in each situation and that this knowledge will directly translate into action.

Naess' work has been influential on radical environmental movements. Terms such as "intrinsic value," and "biocentrism" as well as Naess' work advocating Gandhian nonviolence were picked up by Earth First! in the 1980s. Additionally, Earth First! members and other deep ecologist activists have developed rituals to tap into the experience or intuition of nature discussed by Naess (Taylor 1991: 258–59). Thus, Naess' work is not just an academic theory but has shaped an emerging, lived, worldview.

Inside academia, Sylvan and Bennett aim to enrich the philosophical discussion of Naess' deep ecology, modifying it as necessary to be logically consistent, intellectually convincing, and (hopefully) morally inspiring. Specifically, they reject biospheric egalitarianism in favor of ecoimpartiality, articulate several obligation principles of noninterference, and dismiss Naess' concept of self-realization. Sylvan and Bennett object to the prefix "bio" in Naess' biocentric egalitarianism since it is supposed to encompasses much more than the biotic even though Naess does not adequately explain how his concepts can apply to the inanimate (Naess 1995d: 151–52). They ask how inanimate objects can "live and blossom" let alone have the right to do so (Sylvan and Bennett 1994: 100). They also object to the term "egalitarianism" because Naess does not articulate any criteria for when egalitarianism should be overthrown yet he does overthrow it from time to time (Sylvan and Bennett 1994: 101–2). They fear that without a criterion for applying biospheric egalitarianism humans will end up privileging themselves and deep ecology will surface into a shallow position (Sylvan and Bennett 1994: 102). To attempt to avoid these dangers, Sylvan and Bennett articulate a theory of "eco-impartiality" in which all entities are "objects of value" and "objects of ethical concern"; a smaller group are "objects having well-being, or welfare"; an even smaller group are "preference havers" and "choice makers"; then "rights holders"; then "obligation holders and responsibility bearers"; and finally the smallest group of "contractual obligation makers" (Sylvan and Bennett 1994: 140–42). According to this annular theory, latter traits in the list indicate greater moral standing; any entity that can make contractual obligations would receive greater moral consideration than those without this capacity. Entities within the same category should be treated impartially though they may be treated differently according to need, for example a large active dog may be given more food than a small sedentary one (Sylvan and Bennett 1994: 142, 154).

Their hierarchy of moral concern prompts Sylvan and Bennett to reject Naess' total holism or "relational, total-field image." While they agree that entities in the world are dependent on each other, they object to the easy extension of holism to the idea that there are no dualisms and even more problematically, to the idea that there should be no distinctions between entities. Instead, they argue that individuals and groups deserve moral concern albeit possibly of different sorts, as described above.

To ensure that the annular system does not result in the complete privileging of entities with more morally relevant traits, Sylvan and Bennett follow the Routleys' three obligation principles:

> 1) "not to put others (other preference-havers) into a dispreferred state for no good reason"; 2) "not to jeopardize the wellbeing of natural objects or systems without good reason"; and 3) not to damage or destroy items which "cannot literally be put into dispreferred states ... but can be damaged or destroyed or have their value eroded or impaired."
>
> (Sylvan and Bennett 1994: 147–48)

These principles can be described as "non-interference principles, which exclude[s] unwarranted interference with other preference-havers and unwarranted damage, ill-treatment, or devaluation of items of value" (Sylvan and Bennett 1994: 147). To ensure they do not prohibit eating and the fulfillment of basic needs, Sylvan and Bennett argue that the principle of noninterference only prohibits excessive use of natural items, not all use. With the introduction of these principles, Sylvan and Bennett shift the burden of proof. Instead of requiring people to show that an action would be harmful and therefore prohibited, they instead suggest that "reasons [need] to be given for interfering with the environment" (Sylvan and Bennett 1994: 147).

Sylvan and Bennett also dismiss Naess' notion of self-realization because they think it is chauvinistic. While Naess does include nonliving entities in those that should be cared for so they can "grow and blossom," he certainly does not describe how nonliving or nonsentient entities can be described as flourishing or as self-realizing. Additionally, he frequently uses human-centered and psychological language to describe self-realization yet does not discuss whether or how entities without self-awareness can be self-realized. Sylvan and Bennett do not replace self-realization with any particular principle or metaphysical statement, presumably because they do not see any advantages to the idea, instead focusing on rational justifications of their basic commitments such as intrinsic value (Sylvan and Bennett 1994: 110).

Thus, there are significant differences among those who follow the deep ecology platform. This sort of inconsistency is common in both emerging and long-existing worldviews. If deep ecology continues to be influential, it will likely work out some internal way of reconciling these polarities and/or develop more distinct subgroups. Rather than wait for such a change, I will utilize insights from these various branches of deep ecology in the comparison of ethical systems, recognizing that their diversity can only challenge and improve the identification of the broad principles.

4.2.2 From comparisons to broad principles

Following the criteria for comparison outlined above, this section will move through the ethical principles of Agenda 21 one by one, analyzing when aspects of the three ethical systems support or challenge each principle to discern which broad principles resonate with these systems. To ensure that the ethical systems are considered on their own terms I will also examine unique aspects of each system to determine if additional principles need to be articulated to register the critical points of these ethical systems. This analysis yields the broad principles of responsibility, farsightedness, justice, adequately assessing the situation, adaptability, careful use, and feasible idealism.

Responsibility

A foundation of Agenda 21's normative position is the principle of responsibility. Based on the interdependence of all people, economies, societies, and ecosystems in today's world and the commonsense idea that actions have consequences, Agenda 21 promotes two types of responsibility: admitting that we humans have contributed to environmental destruction and recognizing that we need to change our actions to slow or reverse it (Robinson et al. 1993: 32, 140, 142, 152, 161, 184, 212, 253, 263, 265, 309). Similarly, Nash, Llewellyn, and the deep ecologists assume human responsibility for environmental degradation and to prevent, diminish, or reverse it though they may discuss this in divergent ways grounded on different metaphysical claims.

For example, Nash believed that humans have a responsibility to protect human and biotic rights because people are complicit in environmental destruction and have the ability to act to lessen and, potentially, reverse environmental damage. His virtue of relationality arose from his commitment to responsibility and his assessment of the environmental situation as he maintained that all entities are fundamentally interconnected and that actions by one group may help or harm innumerable others. Relationality as a virtue, according to Nash, requires humans to prioritize interconnectedness and consider the consequences of our actions in and for the whole environment (Nash 1991: 66).

Nash also dedicated a virtue to human care for biodiversity. He thought that unless humans intentionally emphasize the value and preservation of otherkind, our policies will 1) devolve into anthropocentrism that does not appropriately respond to the God-created value in all entities and 2) harm humanity given the interconnectedness of all nature and the unforeseeable consequences of many actions (Nash 1991: 66, 210–14). Thus, he named "biodiversity" a virtue in *Loving Nature* to extend moral consideration to all creatures. Later, he changed this term to "bioresponsibility" to highlight human obligations to biota, rather than merely pointing to the fact of diversity (Nash 2001: 120). It would have been consistent to make a similar move from relationality to responsibility. Regardless of the terms used, for Nash, caring for and being responsible to and for others is considered an outpouring of God's love for humanity and humanity's response in love.

Responsibility is present Llewellyn's work in the idea of *khilāfa*. He maintains that the ontological status of and role for humanity from its very creation, willed by God, includes responsibility to others, primarily humans, though sometimes to animals (Llewellyn 2003: 190). It seems possible that human responsibility through *khilāfa* could be extended to more animals, plants, and, potentially, nonliving entities since all creation has value for Muslims, but this extension has, to my knowledge, not yet been developed.

As Nash's and Llewellyn's assumptions of responsibility arise from their visions of monotheistic creators, we see significant overlap in the way they view responsibility: both see it as a response to God, whether responding to God's love or will. The deep ecologists' vision of responsibility is similar to Nash's in a different respect: their assessment of anthropogenic environmental degradation grounds their idea of responsibility.

Part of the deep ecology platform argues that "Present human interference with the non-human world is excessive, and the situation is rapidly worsening," because human attitudes and use of technology do not recognize the intrinsic value of the world. Thus, they maintain that policies and ideology must be changed. Indeed, they claim that anyone agreeing to these points has an "obligation directly or indirectly to try to implement the necessary changes" (Sylvan and Bennett 1994: 95–99). Certainly this obligation is a way of expressing responsibility.

All three ethical systems stress responsibility for different reasons and have slightly different interpretations of whom one is to be responsible to or for. Yet permeating all of these systems is an admission of human culpability for environmental degradation and a recognition of and demand for change as in Agenda 21. Thus, responsibility holds as a preliminary broad principle after this three-part comparison.

Farsightedness

Though a notion of responsibility is found in many ethical systems, *sustainability* ethics are particularly characterized by their farsightedness, considering the long-range spatial and temporal consequences of actions, policies, and cultural and ecosystem change. Certainly farsightedness was a crucial part of Agenda 21 as emphasized throughout the document in general and in particular studies of aspects of human activity such as agriculture, healthcare, and industry. This principle, in various forms, is also found in the work of Nash, Llewellyn, and deep ecologists and thus functions as a broad principle.

Nash names his future-oriented and long-range virtue "sustainability." For him, it is primarily concerned with long-range intergenerational equity and involves "living within the bounds of the regenerative, absorptive and carrying capacities of the earth, continuously and indefinitely" (Nash 1991: 64). He comes to this virtue through a discussion of "anticipatory rights," the rights of future generations, who will exist as rights holders if we leave them sufficient environmental conditions to come into being. Because Nash takes scientific assessment seriously, he knows that past and present human activity threaten the ability of future humans to meet their basic needs as they threaten ecosystem services (Nash 1991: 206–8). Thus, he claims that acting to preserve the environment today for the long run is the best way to protect the rights of future generations. Nash knows that balancing the needs of the future and present may be quite difficult since they may suggest different courses of action, but does not think it is an insurmountable challenge since many "behavioral patterns – like sustainability and frugality – that will benefit future generations will also benefit the present one" (Nash 1991: 209).

Nash's use of sustainability is very similar to that of Agenda 21. Both require many other norms for their proper fulfillment and suggest a goal of an ideal state in which natural processes can continue indefinitely. Yet to avoid confusion between the multiple meanings of sustainability in Nash's work, I suggest differentiating between the goal of sustainability, "living within the bounds of the regenerative, absorptive and carrying capacities of the earth, continuously and indefinitely" (Nash 1991: 64), and the principle of farsightedness, in which it is right to consider the long-range spatial and temporal consequences of decisions, policies, actions, and ecosystem change. Making this move enables us to acknowledge the interrelationship of many norms in sustainability discussions, avoiding the facile oversimplification that valuing the future is the only normative claim of sustainability (Nash 1991: 64, Sylvan and Bennett 1994: 126, 172).

Llewellyn does not articulate a specific principle directing people to consider the future from within Islamic law but clearly recognizes that something like farsightedness is a critical component of an Islamic environmental law as he focuses upon legal instruments including *waqf*, *ḥimā*, and *ḥarām* which have historically and could today be used to prioritize the preservation of natural resources for the long run. A *ḥimā* is a piece of land set aside to follow God's purposes by serving the economic and environmental good of the whole community. Additionally, a *ḥimā* must have more benefits than drawbacks to society as a whole and cannot limit local people's access to resources that fulfill their basic needs. Building on or commercializing the land of a *ḥimā* is prohibited though it is sometimes able to be used for grazing, cutting trees, or making honey at regulated seasons and rates to ensure that the land is preserved for future generations (Haq 2003: 144, Llewellyn 2003: 212). While *ḥimās* have been most often used in these traditional ways, technically, they could be established for any reason that meets the criteria. Consequently, many Islamic environmentalists think *ḥimās* promote land and species preservation, improve water supplies, or serve as recreational, research, or educational areas to promote understanding and appreciation of the environment (Llewellyn 2003: 215). Indeed, Llewellyn claims that *ḥimās* are "the most important legal instrument in the *Shari'a* for conservation of biological diversity" (Llewellyn 2003: 216). Though *ḥimās* are an ecologically promising idea, their numbers have diminished significantly from approximately 3000 in Saudi Arabia alone in 1965 to only a few dozen today as lands once under tribal authority became nationalized and populations have risen (Llewellyn 2003: 213–15). Thus, Muslims working to promote *ḥimās* for environmental reasons will have to overcome prevailing political trends while reenvisioning how *ḥimās* will be governed and for what purposes (Llewellyn 2003: 217).

In order to motivate people to designate resources for *ḥimā* Llewellyn and others promote local education and the revitalization of *ḥarām* laws. A *ḥarām* is a "sacred territory, inviolable zone, [or] sanctuary" used to promote the welfare of all inhabitants (Haq 2003: 144, Llewellyn 2003: 208, 210). They are areas similar to a greenbelt surrounding each Islamic settlement and natural and developed water sources. *Ḥarīm* around settlements were traditionally used for forage and firewood but could also be used to preserve species intentionally, cleanse the air, and provide green space for recreation or aesthetic purposes. *Ḥarīm* around water also prevent water pollution, facilitate the maintenance of water sources, and, by prohibiting new wells within their boundaries, preserve the water supply of existing wells (Llewellyn 2003: 210–11). While the use of ḥarīm to protect water sources is in jeopardy today because "the municipal commons of settlements are presently overexploited and not managed; the inviolable zones of water sources are largely ignored," Llewellyn and others argue that revitalizing this preservation and future-oriented part of Islamic law could significantly impact and the environment of future generations (Llewellyn 2003: 211).

The legal traditions of land preservation may be helpful in creating an Islamic environmental ethic, but their decline in recent years makes it important to look for other environmental resources within Islamic law. One of the most promising is the *waqf*, or charitable trust. *Waqfs* are established by a benefactor for the good of the community and have historically been a major source of funds for institutions such as hospitals and schools in Islamic societies. *Waqfs* can also support land or water sources set aside for community well-being as illustrated by the actions of Othman, later the third caliph, who bought the well of Ruma and turned it into a *waqf* for the good of the people at the Prophet's advice (Faruqui 2001: 2). By establishing more *waqfs* to preserve environmental assets today, land could be preserved for future generations as they slowly demonstrate the benefit of environmental *waqfs* and encourage others to donate such lands for preservation. As Islamic legal experts advocate the use of ḥimā, ḥarīm, and *waqfs* they demonstrate their commitment to farsightedness.

Farsightedness also plays a role in deep ecology. It is most noticeable in their critiques of standard economic and political practices that ignore the long-term denigration of natural resources and ecosystems (Sylvan and Bennett 1994: 126, 172). Aside from these discussions of economics, Naess, Sylvan, and Bennett tend to focus on the present because they want people to start changing their attitudes about intrinsic value and acting accordingly right now. Thus their work is not as explicitly farsighted even though much of their moral outrage is based on their desire for biota to continue into the far future.

Consequently, with farsightedness we see a principle resonant with each system whether explicitly discussed under another name (Nash), emphasized in the types of legal instruments highlighted (Llewellyn), or underlying and presumed in their recommendations (deep ecology).

From equity to justice

Through their farsightedness, sustainability ethicists look toward a certain type of future: one centered on equity, or more broadly speaking, justice. Agenda 21 emphasizes the ethical principle of equity, usually an equitable ability to meet basic needs, in its general pronouncements and its claims that often-marginalized people including women, children, the poor, citizens of developing countries, and indigenous people are equal to all others and should be treated as such (Robinson et al. 1993: 26, 46, 50). Yet it does not explicitly discuss the very real problems of racial or ethnic environmental injustices; focuses on individuals, rather than communities; and does not specifically discuss equity with respect to nonhumans, three issues many find problematic. Thus, the idea of equity needs to be extended into a broad principle acceptable to the ethical positions in our comparison (Bullard 1994a: 59–61, Heredia 1994: 123–27, Heyd 1994: 131, Ott 1994: 219, Paden 1994: 261–63, Rolston III 1994: 270–80, Warren 1994: 321, Weiss 1994: 362).

Nash articulates a virtue of equity that occurs when 1) goods and services are distributed so that every person can participate in society with dignity and 2) the negative effects of human activities such as pollution are distributed such that no person or group is harmed disproportionally to the benefits they experience from the same activities (Nash 1991, 65). When equity is ensured all will have their basic rights, assuming there are enough natural resources to go around.

Llewellyn highlights similar ideas as he stresses the ideals of prioritizing universal needs, assisting those with less power, and prohibiting harm. Similarly, Islamic studies of water distribution highlight concerns for equity. Water, as one part of creation, is seen as a gift from God, owned by no person. This conviction, and the importance of water for all life, caused the Prophet to discourage the sale of water and forbid the sale of excess water. Also, as noted, the Prophet motivated Othman to establish the well at Ruma as a *waqf* and give away its water for free, emphasizing the needs and rights of all people to water (Faruqui et al. 2001: 12). Yet as Llewellyn and others focus on the good of the community as a whole, rights-based language is not always sufficient. A broad principle to resonate with these ideas must prioritize the ability of all individuals and communities to meet their needs equitably.

The principle also needs to be extended to account for Nash's commitment to biotic rights and the deep ecologists' commitment to intrinsic value. Nash separates equity among humans from biodiversity because he thinks that humans have some rights in addition to those of otherkind and thus deserve special, not equal, treatment under certain conditions. He also separates these virtues to call attention to biotic rights which have too often been ignored. Similarly, Naess' commitment to biocentric egalitarianism and Sylvan and Bennett's commitment to ecoimpartiality call for an ethical principle not merely focused on humanity.

Given these variations in visions of equity or justice more reflection is needed to articulate a broad principle to encompass these views. While two different principles, for humans and all others, could be articulated, as Nash does, the emphasis on biotic rights would not be acceptable to all and the sharp division between humans and others is not acceptable to the deep ecologists. Thus, this is a time to capitalize on the ability of the broad principles to encompass different, possibly contradictory specifications. I suggest a broad principle of justice which is minimally specified by equitable distribution of goods and services and may also include some combination of participation in decision-making among humans, the consideration of intrinsic value for all or biotic rights or moral concern for groups of entities that share morally relevant traits. This move will allow such specifications without requiring them, enabling more people to use the broad principles while still holding onto their deeply held beliefs. Despite this broadness, the principle still can provide traction for ethical decision-making about index development, as will be illustrated in Chapter 5.

Adequate data assessment

Responsibility, farsightedness, and justice are all connected to the way people understand the state of society and the environment. I argue that adequate data assessment is an ethical principle in Agenda 21 because its authors presume that normative goals for sustainability are not just abstract ideals, but emerge out of the ecosystemic and societal context in which the authors find themselves. Thus, to develop and implement an action plan for sustainable development requires that communities and nations understand the environmental, economic, and social details of the situation in which movement toward sustainability is desired. To refuse to acknowledge, or seek out, the best knowledge about an environmental situation and its relationship to human society is to act unethically. Nash, Llewellyn, and deep ecology all rely on something akin to adequate assessment of the situation though they add specifications such as theological, philosophical, and intuitional data beyond the scientifically measurable data that Agenda 21 focuses on. Thus, adequate assessment of the situation works as a broad principle as long as it is expanded beyond the specifications pertinent to Agenda 21.

The recognition of the intrinsic value of otherkind is a particular way of specifying the broad principle of adequate assessment of the situation, though not one emphasized in Agenda 21. Deep ecologists believe that they gain this knowledge of the world through intuition or feeling, coupled with rationality, and understand it as a definite property of nature that must be considered in any moral decision just as pollution rates or economic factors are. Indeed, for some deep ecologists, intrinsic value may be more important than economic or political concerns; scientific data may be used primarily to assess the status of entities with intrinsic value.

Similarly, Nash's virtue of relationality relies on recognition of the interdependence of humanity and the environment, a type of assessment of the situation. While it is understandable that Nash would want to emphasize relationality given its neglect in the past, relationality could be subsumed under the broad principle of adequate assessment of the situation and responsibility. This move classifies mandates about observations in one principle while enabling the moral imperative of responsibility to stand on its own as a broad principle. Dividing the principles in this way enables both responsibility and adequate assessment of the situation to resonate with more traditions.

Adequate assessment of the situation also encompasses Nash's idea of the virtue of humility, a realistic assessment of the strengths and weaknesses of oneself, humanity, and the environment at large. Contrary to some popular ideas, Nash's humility is not perpetual self-denigration and does not require relinquishing one's talents, needs, or ideals, but rather involves 1) recognizing when one does not currently have the proper information or power to make and act on a decision; acknowledging that some information may not be obtainable because existing theories or instruments are limited or because complex systems make certain facts or relationships indeterminable; and 2) understanding, to the best of one's ability, the restorative capacities of the environment (Nash 1991: 66–67, 156–57). Humility prompts Nash to advocate for multidisciplinary approaches to environmental problems, recognizing that no one person or discipline is sufficient to address environmental problems. Thus, it too is a specification of a realistic assessment of the situation that goes beyond Agenda 21's focus on knowledge acquisition to recognize the limits of knowledge and action.

Building from the appropriate assessment of our own capabilities stressed in the virtue of humility, Nash also advocates a careful assessment of proposed solutions to environmental crises through the virtue of sufficiency. According to this virtue, one is ethical if one proposes and carries out solutions to environmental problems that are sufficient to address ecological concerns given the physical situation, political climate, knowledge, and technological and moral capacities (Nash 1991: 66, Chapter 8). Agenda 21 certainly aimed to prioritize sufficient solutions as it prompted developing science, educational, business, and government solutions to sustainability issues while attending to cultural values. It did not, however, examine moral sufficiency nor did it address the limits of these capacities to the extent Nash was able to do given his focus on humility. Thus, Nash's virtues of humility and sufficiency are not only intimately related to adequate assessment of the situation but also expand the set of specifications that can be classed under it.

Nash's virtue of solidarity aims to ensure that strategies for solving environmental problems are created in the community best equipped to solve them (Nash 1991: 65–66, 215–21). Insofar as solidarity is intended to identify the proper group to enact a particular solution to environmental problems it seems to be a part of adequate assessment of the situation.

Nash and Llewellyn also rely on theological assessments of the world including the belief that since God created all, all are valuable and, in Llewellyn's case, the idea of God making people *khilāfa*. While Agenda 21 does not necessarily advocate these particular approaches, it does emphasize developing strategies that rely on the values of individual communities.

Each of the three systems also relies on contemporary science in its assessments of the situations. Nash, for example, relies on scientific data about specific environmental problems including pollution, ozone depletion, and global warming to begin his analysis (Nash 1991: 23–29). Llewellyn advocates using a variety of sciences to adjust boundaries for *ḥarīm* around water sources to protect today's deeper wells from new contaminants (Llewellyn 2003: 211). Deep ecologists also use scientific assessments to ground their claims that something is wrong with contemporary interactions with the environment. They do, however, tend to be a bit more wary of technological fixes than Nash or Llewellyn and certainly less than Agenda 21. Such hesitancy arises because they think science and technology have been significant factors leading to environmental crises and because some, especially Naess, emphasize self-realization and intuition as methods of interacting with and understanding the world. Despite these variations in the use of scientific assessments, Nash, Llewellyn, and the deep ecologists all use scientific assessments as a part of their multifaceted assessments of the world.

In sum, there are many ways of assessing the situation beyond those emphasized in Agenda 21. Scientific, theological, and intuition-based assessments, the recognition of intrinsic value and relationality, and the need to be humble and have sufficient solutions can all be subsumed under a broad principle entitled "adequate assessment of the situation" as long as one realizes that "assessment" is not limited to the results of modern science.

Adaptability

Similarly, adaptability is found in all three ethical systems though Nash and Llewellyn emphasize it more than deep ecologists do. Recall that Agenda 21 prioritizes adaptability of environmental law with respect to the level at which a policy is implemented, the degree of interaction with various cultural values, the uneven resources of political entities, and the need to change policies over time. It arises out of the principle of assessment of the situation and guards against the hubris of believing that one knows enough about environmental situations to make a policy decision for all time and all situations. Adaptability also indicates something about the ethical method assumed throughout Agenda 21: general ethical terms are implied with the understanding that they can be implemented by adherents of widely diverse cultures and that ethical traditions may need to be revised in light of critical reflection, new situations, and new knowledge. Thus, as with adequate assessment of the situation, we see that adaptability imparts ethical content (act and make decisions in ways that can be applicable to many contexts) and ethical method. Adaptability as a broad principle encompasses Nash's virtue of adaptability, methods of Islamic law, and, to a lesser extent, the variations allowed beyond the platform of deep ecology.

Adaptability, explicitly named as one of Nash's environmental virtues, responds to the fact of limits. Adaptability dovetails with humility and sufficiency as it requires that plans for the future include appropriate contingency plans in case of uncertainty whether due to natural disaster, human error, or the limits of knowledge and plans that are flexible enough to respond to new challenges and opportunities. While adaptability has not always been considered a virtue, Nash argues that it is essential for an ecological age because humans' inability to adapt to the changing environmental circumstances has led to many of the ecological crises we are facing today (Nash 1991: 64). Thus, Nash advocates adapting principles of environmental ethics, policy recommendations, and actions to changing environmental circumstances, needs of the human population and political and economic climate (Nash 1991: especially chapters 1, 2, 8). In this way his sense of adaptability is nearly identical to that of Agenda 21; the main difference is that Nash describes the necessary changes to ethics in much more detail.

Llewellyn's work resonates with adaptability in his discussion of *qiyās*. *Qiyās* is a process of analogical reasoning developed in classical Islamic law to make decisions about issues for which there was no legal precedent in the *Qur'an* or *ḥadīth* that can be used to make ancient Islamic laws relevant to contemporary environmental dilemmas. For example, in traditional Islamic law, one was responsible for potential harm to the community resulting from incidents on one's own land. Thus, the walls near the edge of one's land needed to be well maintained so they would not crumble and fall onto another's land or harm people or property. From this law, Muslim scholars have begun to argue that one is responsible for any long-term effects from actions on one's property, for example, the release of pollutants into the air. This ability to extend old laws to new situations is critical for an environmental ethic as environmental disruption has spatial and temporal consequences to a degree previously unstudied and unknown.

Naess recognized that deep ecology is adaptable to different ecosophies and therefore that various groups of deep ecologists may emphasize different specifications of the basic platform (Naess 1985, 226). For instance, Sylvan and Bennett encourage people to adhere to the basic framework of deep green theory while adapting it to their particular situation (Sylvan and Bennett 1994: 175). They also acknowledge that as new data is accumulated value judgments and interpretations of previous experiences may need to be adjusted. Thus, their system is adaptable according to the ethical backgrounds of various deep green theorists, the physical situation of entities of moral consideration, and the accumulation of new information (Sylvan and Bennett 1994: 144).

An expanded version of adaptability to encompass changing ethics, as well as variations in policies, cultural values, and economic circumstances, will encompass the types of adaptability highlighted in Agenda 21 as well as the work of Nash, Llewellyn, and the deep ecologists.

From efficiency to careful use

Given the limited number of resources to be distributed with equity and the rampant pollution and environmental disruption associated with human activity, many argue that environmental resources should be used in a cautious way. Agenda 21 encourages the efficient allocation and use of natural resources as well as, to a lesser degree, decreased consumption. Yet the principle, tentatively named efficiency after examining Agenda 21, requires broadening and a new name to encompass the priorities of others. Then it can encompass Agenda 21's focus on technological fixes of efficiency as well as the emphasis on ideological changes, decreases in consumption, frugality, prohibitions against waste, and nonmaterial quality of life of Nash, Llewellyn, and deep ecologists.

For instance, Nash argues that frugality is necessary for a complete ethic. Indeed, Nash probably spent more time exploring the meaning of and need for frugality than any other virtue (Nash 1994, Nash 1995). He is adament that frugality does not just entail giving up lavish lifestyles; it also implies efficiency, recycling, durability, reparability, conservation, restrained consumption, and "a satisfaction with material sufficiency" (Nash 1995: 144). Nash argues that while frugality has been deemphasized in recent years it needs to be reclaimed to be continuous with Christian economic ethics and to respond to the tension between economic systems that depend on continual growth and the finiteness of the natural world (Nash 1995: 137–39, 158). For Christians, frugality is rooted in Jesus' critiques of the idolatry of wealth in order to focus on what is more important: affirming Christian belief in the connection with and dependence of humanity on God as creator, incarnate Christ, and "sacramentally present Spirit" (Nash 1995: 151–52). Frugality is also grounded in biblical concerns for "covenantal justice," which focuses on the preferential treatment of the poor so that they all have their basic needs fulfilled (Nash 1995: 152). To ensure that frugality supports justice, Nash emphasizes that "frugality is a progressive virtue"; it is more important the more one has (Nash 1991: 65, 183). In other words, frugality should not limit people's access to the resources necessary for well-being and participation in society.

Similarly, Llewellyn recognizes prohibitions against waste in traditional Islamic law firmly rooted in the *Qur'an*. *Suras* 6:141 and 7:31 state "Do not waste: verily He loves not the Wasteful!" From this and related *suras*, some Muslims, including Llewellyn, conclude that all types of "wasteful overconsumption and extravagance" are forbidden by the *Qur'an* (Llewellyn 2003: 199). These conclusions are supported by aspects of Islamic law which stress that consumption should be limited to avoid ostentatious acts and to allow all to have resources necessary to live. For instance, there is a prohibition against hording water beyond what one needs for one's own immediate use (Llewellyn 2003: 205). Furthermore, specific *ḥadīth* prohibit wasting water. One of the most famous of these *ḥadīth* describes the Prophet rebuking a Muslim extravagantly using water while washing before daily prayers, stating that one can be wasteful "even if you perform them [ablutions] on the bank of a rushing river" (Ozdemir 2003: 14). From the claim that waste is possible when performing ritual cleansing in the face of abundant resources, Llewellyn and other scholars infer that wasting water is not permitted during any activity. Additionally, they suggest that such prohibitions could be extended to other natural resources such as minerals, fossil fuels, or clean air (Llewellyn 2003: 237).

Comparative ethics for sustainability 107

The deep ecology platform also promotes reducing consumption and conserving resources. It clearly distinguishes between quality of life and quantity of goods consumed, noting that quality of life, not quantity of consumption, is its major goal. Additionally, one of its platform principles establishes that actions permitted to fulfill needs may be unacceptable in the pursuit of mere wants (Naess 1995a: 68). While Sylvan and Bennett certainly promote reducing consumption and limiting wants, the efficiency component of careful use receives much less emphasis in their work, probably because they are wary of technological optimism and thus hesitate to endorse many technical solutions (Sylvan and Bennett 1994: 218).

The deep ecology position on quality of life, Llewellyn's emphasis on conservation and prohibitions of waste, and Nash's position on frugality differ enough from Agenda 21's focus on efficiency such that none of these terms can function as a broad principle for this constellation of ideas. Efficiency through technological fixes does not necessarily register Nash and Llewellyn's focus on God or the focus of Nash, Llewellyn, and deep ecologists on non-material basic needs (Nash 1994: 55). Furthermore, neither frugality nor efficiency necessarily encompasses the other – someone who is frugal may be efficient in his or her use of money but not in his or her use of material resources or time. Similarly, someone may efficiently produce goods that promote careless spending. For example, consider the recent trend toward disposable cleaning tools such as single-use mopheads and towelettes embedded with cleansing chemicals. All of these products could be produced very efficiently and yet since they are intended to be discarded right after use they increase the total amount of resources used. Thus, a new term beyond frugality and efficiency is needed to name the broad principle that encompasses these terms.

"Careful use" is a good candidate because it can encompass a number of concepts. Certainly, being careful in one's use of resources can imply being cautious with them and discouraging waste, a part of efficiency. Think of the child first allowed to pour milk into a glass. What does the parent say? "Be careful." The child needs to take care not to spill – because to spill is to waste milk and to use more cleaning products, time, and a parent's patience than is necessary. Thus, careful use can discourage waste. Careful use can also be specified by only using resources when needed, the other component of efficiency. As lost hikers carefully ration their food so too can governments, companies, and individuals carefully use products and chemicals. Careful use can also be specified in the frugal sense of avoiding the seduction of affluence so that what is most important, in Nash's case, relationships with God, people, and the world, are not neglected. This specification of careful use also resonates with the many systems of religious economic ethics that understand that the desire for material possessions leads to a false sense of fulfillment. Finally, careful use encompasses recycling and minimizing waste streams, for to do so is to be careful and deliberate in one's actions so that what one cares about can be preserved or obtained. Thus, careful use is a good candidate for a broad category in that it can register multiple divergent ideas including frugality, efficiency, and recycling.

Admittedly, some will critique the broad principle of careful use because it includes the term "use"; critics will argue that since the misuse of natural resources is what led to environmental destruction in the first place we should not encourage people to use the environment. Certainly, an instrumental view of the environment has been and can be problematic, but pretending that people do not need to use resources to stay alive is completely unrealistic. Implementing "careful use" with the other principles as these interrelations demand will guard against distorted interpretations of use that view all entities merely as a means to an end.

Cooperation

The last ethical principle I identified in Agenda 21 is cooperation. As you may recall, however, I questioned whether it was properly an ethical principle as it is a condition presumed in the activity of social ethics and therefore could be subsumed under a broad notion of adequate assessment of the situation. Cooperation enters into sustainability discussions when people assert that humans have a duty to cooperate with others as they work toward sustainable development. After all, the dictatorial imposition of measures to achieve sustainable development, an extreme alternative to cooperation, would counter Agenda 21's emphasis on the rights of individuals to participate in decision-making processes according to their cultural, religious, ethical, and personal values and its commitment to a diversity of values in general (Robinson et al. 1993: 2, 26, 28, 42, 45, 47, 51–55, 57, 68, 72, 75, 492–95, 620–26). Nash, Llewellyn, and deep ecologists value cooperation to at least some degree. Yet an examination of their systems and a reflection on cooperation's relationship to the other principles suggests that it should not have its own place as a broad principle.

For instance, humility contributes to Nash's recommendations for cooperation among all peoples regarding environmental concerns because he realizes that the scope of contemporary environmental problems entails that people from any one discipline, nation, or socioeconomic class cannot fix these problems on their own. Rather, they must humbly admit that each can contribute to the solution and that they do need each other. Similarly, Llewellyn calls for the collaboration of "socio-economists, planners and local communities" to determine new *ḥarām* boundaries around wells (Llewellyn 2003: 211). Llewellyn's focus on Islamic law also leads him to focus on cooperation within the Islamic community whether in the care for the poor, in the development of laws within schools, and by *mujitahid* for a community. Naess focuses on the cooperation of individuals in local communities because he believes this is the best scale for environmental action because locals are best able to account for their specific environmental conditions and needs. Thus, he advocates for the decentralization of governments and all services (Naess 1995d: 152–53). While Sylvan and Bennett focus on encouraging each person or group to spread the message of deep green theory using their own skill set they also argue for both top-down and bottom-up approaches to the spread of deep green theories and emphasize the need for teachers, advertisers, unions, churches, corporations, and governments to spread the word about environmental ethics (Sylvan and Bennett 1994: 211–26). Thus cooperation is emphasized to some degree in each of these ethical systems.

While one could potentially name cooperation as a broad principle to encompass these various ways of responding to environmental problems, it seems to be a variation on adequate assessment and feasible idealism (to be discussed below). Instead of merely assessing the situation now, mandating cooperation entails an assessment, in consultation with the other ethical principles, of who needs to be involved to address the situation's needs. This assessment is a logical expansion of the broad principle of adequate assessment of the situation. It, like sufficiency, which is already assumed under adequate assessment, is focused on predictive, future-oriented assessment of the situation, on possibilities that are feasible yet are aiming toward ideals. To avoid duplicating principles, I suggest subsuming cooperation under these other principles.

Feasible idealism

The ethical principles identified in Agenda 21 have provided a basis for our comparison which yields broad principles. This method, however, is not a sufficient way to undertake such a comparison because it does not adequately examine the various ethical systems on their own terms. To ensure that elements of each ethical system examined are considered on their own terms, as is needed in a rigorous comparison, we must assess whether they suggest any ethical principles not discussed thus far. Indeed, when examining these systems of ethics another principle emerges: feasible idealism.

Feasible idealism is most explicit in Nash's work. As we have seen, Nash's virtues are supported by his theology of love and attention to a revised natural law that attends to embodiment and knowledge from the natural and social sciences. A careful look at Nash's work also reveals a subtle mix of idealism and attention to the feasibility of actions suggested by his ethics. As one who often wrote about the role of churches and Christians in the political sphere, Nash was well acquainted with what he deemed "the politicians' moral dilemma": how to be elected and have political influence while adhering to one's values (Nash 1996: 8). Nash describes this balancing act as seeking a middle ground between perfectionism and accommodationism. For Nash, perfectionists are unwilling to compromise in the quest for their ultimate ideals. Often this means that they have little power or influence on decision-making. On the other hand, accommodationists are those who see power as an end in and of itself. They are less attached to any one system of values yet because they are willing to accommodate many points of view are often able to stay in power longer (Nash 1996: 8–9). Nash argues that decision-makers ought to occupy a space between these two poles: they must follow their ethical ideals and know that implementing them will involve some compromises to realize "the best possible" (Nash 1996: 10).

Nash understands that this ethical task will be supremely difficult, especially in the case of frugality and other virtues that run counter to the dominant culture. Frugality, for instance, simply does not fit the ethos of the "Sumptuous Society" of contemporary America. Rather, it demands a new economic paradigm because it acknowledges the biophysical limits of the planet. Ethicists must also be careful when advocating frugality because a sudden wide-scale shift to the new economic paradigm demanded by frugality could cause a massive disruption in society and significant harms to the most vulnerable (Nash 1994: 57–59). Because moral norms such as frugality will meet with popular resistance and may have negative effects if implemented suddenly, Nash recognizes that decision-makers must pay attention to the feasibility of implementing norms as well as their short and long-term implications.

Thus it seems that Nash follows another virtue in addition to those he explicitly names: mediating between the ideal and the possible. People with this virtue acknowledge challenges to environmental action, yet have hope and focus their attention on effective, feasible actions (Nash 1989: 43, Nash 1991: 206). Of course, hope may be grounded in a number of sources including the power of God, logical reasoning about what is possible, and technological optimism. Balancing the ideal and possible is particularly important for the actual application of ethics to concrete problems; the more specific an ethical dilemma is the more this norm will come into play.

Thus, it is not surprising that similar ideas most significantly and explicitly arise in Llewellyn's work when he outlines the actions necessary to foster and implement the discipline of Islamic environmental law. He argues that jurists would ideally be schooled in the methods and substantive rulings of Islamic law, international law, and the social and natural sciences pertaining to environmental and social welfare. Since such interdisciplinary programs and the professors to teach them are few and far between, Llewellyn encourages team-taught classrooms using legal and environmental studies professors, or double majors in law and environmental sciences until integrated environmental law programs are established (Llewellyn 2003: 238–39). Here Llewellyn articulates an ideal as well as feasible steps toward the ideal given contemporary social contexts.

Llewellyn also walks the line between idealism and feasibility as he describes the process of creating Islamic environmental laws. He notes that once the basic principles and norms of Islamic environmental law are delineated there will be considerable latitude in the way these ideas can be implemented. After all, the law is always supposed to be influenced by the needs of the society and the environment in which it is made (Llewellyn 2003: 208, 230). For example, the needs of the people and the characteristics of their ecosystem will shape the sorts of laws about well protection and farming devised with Islamic law. Furthermore, Llewellyn knows that conservation efforts must be economically viable in the short and long term in order to be just and implementable.

Sometimes Llewellyn tends toward idealism. He recognizes that environmental jurists will not be influential without substantial changes to the legal system since *Shari'a* courts often have no jurisdiction over environmental issues, yet he does not suggest how such courts could become more influential (Llewellyn 2003: 239). Because Llewellyn posits that Islamic law is the best Islamic resource to combat environmental destruction a more significant discussion of the implementation of Islamic environmental law in the contemporary world would greatly strengthen his argument.

Sylvan and Bennett's *A Greening of Ethics* reveals their commitment to the ideals of deep green theory. They believe that eliminating chauvinism, the central goal of their ethic, can resolve environmental crises and social problems of all sorts. A first glance at their work would also indicate that they are as committed to assessing the feasibility of their ethical project as are Nash and Llewellyn. After all, they reflect on the marketability of ethics in general in order to determine how best to promote their own ideas (Sylvan and Bennett 1994: 177–204, 214–15). They also identify the particular sorts of individuals, small groups, and major organizations that will be most effective at spreading their vision (Sylvan and Bennett 1994: 211–26). Yet while they have elements of idealism and feasibility in their work, they are closer to idealism than either Nash or Llewellyn because they do not adequately assess the feasibility of their proposals. In particular, they do not recognize the pervasiveness and strengths of the moral, economic, and social systems they oppose or the possibility that their action plans may have limits or may be implemented imperfectly.

Feasible idealism is also illustrated in the work of Earth First! members who follow ideas of deep ecology. Though they often describe other subgroups as moral failures (for failing to remain true to biocentrism by focusing on social justice or by failing to recognize that social justice is a part of biocentrism) rather than framing their discussion in terms of feasibility, Bron Taylor's analysis of their beliefs and actions indicates that their decision about using "civil disobedience or ecotage" in their protests is "more about strategy and tactics than fundamental moral differences: both factions remain biocentric" (Taylor 1991: 263–64). Such analysis suggests that though the deep ecology activists are highly committed to ideals, they also significantly attend to the feasibility of their strategies, as influenced by their assessment of the situation.

In these analyses, we see that deep ecologists tend toward idealism while Llewellyn stresses the social acceptability of new laws and Nash focuses on the political feasibility of environmental norms. These differences arise because Sylvan and Bennett focus on theory more than application while Llewellyn promotes Islamic law and Nash works with the secular American legal system. Despite these differences, they all recognize, at least to some extent, that when they implement their environmental ethic, they need to balance their ideals with the feasibility of implementing them. For these reasons, I argue for the creation of a new broad principle to encompass this dimension of their work.

I've struggled to find a decent term for this principle but think that "feasible idealism" works fairly well.[1] One might be tempted to just name it feasibility, but none of these authors go so far as to say that one should only do what is feasible, they all desire to push the feasible toward their ideals. Thus with this hybrid term I aim to indicate that the principle does not involve taking the easy way out but rather, makes concessions where necessary, pushing toward the ideal.

Interestingly enough, after defining this principle, it is recognizable in Agenda 21 as well (Robinson et al. 1993: xxv). For the most part, this principle is inherent in the process by which Agenda 21 was written and endorsed. After all, achieving the consensus of over 170 nations will serve as a deterrent to the most idealistic and radical propositions conceivable for an action plan. Sometimes, this attention to political feasibility means that Agenda 21 is less specific and visionary than many would hope. For example, disagreements among oil producing nations prevented Agenda 21 from articulating a stronger energy policy (Robinson et al. 1993: xxv). The participation of many nations did, however, contribute to Agenda 21's awareness of the diversity of cultural contexts. Agenda 21 is particularly conscious of the monetary, technical, and knowledge-based assistance developing nations will need to implement its plans, and the ways in which local moral systems will influence population control policies (Robinson et al. 1993: 2, 20, 24, 28, 34, 42, 45, 47, 51–55, 57, 68, 72, 75, 494–95). In this way, feasibility can promote local ideals.

Certainly, there may be other candidates for a broad principle that emerge through additional comparisons or in light of new knowledge or situations. For now, however, the working list consists of responsibility, farsightedness, justice, adequately assessing the situation, adaptability, careful use, and feasible idealism which together register the significant elements of the ethical systems of Nash, Llewellyn, and the deep ecologists studied as well as that of Agenda 21. This list represents a broadening of Agenda 21's responsibility, farsightedness, and adaptability; a broadening of adequate assessment of the situation to include various theological and metaphysical claims as well as cooperation, originally proposed as a principle in its own right; the replacement of equity and efficiency with careful use; and the identification of a new principle, feasible idealism. In addition to identifying the principles, this process of comparison illustrates the way in which comparisons can yield and revise vague principles for collaborative decision-making.

4.2.3 Broad principles of sustainability ethics – a working list

Since the above comparison to yield the broad principles may be more detailed than all readers will desire, and because the process of comparison can overshadow its results, this section includes a short summary of each principle. Remember that these principles are an interrelated set in that the application of each is influenced by the others. Additionally, the principles are broad in that they can be specified in a variety of different ways and yet provide bounds for decision-making that helpfully constrain the acceptable decisions and therefore can influence index development, as indicated in the chapters to come.

An *adequate assessment of the situation* entails the honest, accurate assessment of the situation in which an ethical decision or action is to be made. The situation can consist of several elements: the world as it exists apart from humanity; the world in relation to humanity; and the actions and capabilities of humanity as a whole, in groups, and as individuals. It necessarily relies upon metaphysical and ontological assumptions about the world, for example, whether it has intrinsic value. The identification of these categories is not intended to indicate that they are ontologically distinct. Rather, being explicit about the three ensures none is neglected. An adequate assessment of the situation also registers ways of responding to the situation including cooperation and virtues like humility and pride that together promote the balance between realizing one's limits and using one's strengths (Nash 1991: 66–67). This principle also encompasses evaluations of who can best address an issue and the level of cooperation needed to do so. Consequently, it encourages attentiveness to scientific knowledge about ecosystems, industrial processes, pollution, and human needs; sociological, psychological, and religious knowledge about the imperfections of people that necessitate laws, punishment, and forgiveness; and ethical knowledge needed to adhere as closely as possible to our ideals as specified in this and other principles. Admittedly, the term "adequate" in adequate assessment of the situation deserves some reflection. Truly, it is difficult to determine exactly how much analysis is sufficient. However, this broad principle encourages people to consider all of the sorts of data and methods they find relevant and discover what this information indicates about the situation at hand. Over time, these data and methods can be assessed for adequacy based on how well they mesh with experience and prevent environmental degradation. An assessment of this situation is likely to be adequate if it was compiled in consultation with people representing diverse disciplines and worldviews, and if it is broad in scope, yet attentive to details. Thus, adequate assessment, when applied to the development of sustainability indexes and indicators, suggests that such development should follow the best available guidelines for such activities: the polarities of technical rigor and accessibility, comprehensiveness and ease of use, and ideal and actual data.

Adaptability encourages us to adapt ourselves and our policies to the world and our knowledge of it, both of which continually change. I recommend that adaptability includes incorporating precaution into our policies given the inherent limits of our knowledge of the world and uncertainties about its changing conditions. Finally, adaptability can also entail sensitivity to the particular cultural and environmental contexts in which we find ourselves. This principle is separated from adequate assessment of the situation because we humans need to be continually reminded of the dynamic nature of the world to avoid clinging to one approach or one bit of knowledge. Otherwise, we may maintain an approach even when it does not work in a particular situation.

Justice encompasses the equitable distribution of resources including the physical (food, air, clothing, and shelter), the mental (fulfilling work and security) and political (the ability to participate in the process of making decisions that will affect one). It is an excellent example of how the other principles are needed to specify the application of any one principle. In particular, the extent of "justice for all" must be determined through the use of the other principles. As will be discussed in more detail in Chapter 5, contemporary assessments of the effects of environmental burdens and benefits on human society indicate that these effects are not distributed equitably – people of color and the poor are disproportionately burdened by environmental degradation though they are less likely to have benefitted from the activities that cause such environmental degradation or to have been involved in the decision-making process that led to such activities. Recognizing this state of the situation, claiming responsibility for the effects of one's decisions and actions, and valuing justice, those who prioritize justice for all are increasingly focusing on the plight of the least advantaged in society, arguing that if their physical, mental, and political needs are met then the needs of *all* people will be met.

Of course, environmental thought does not only focus on humanity. Indeed, some of the most controversial debates in environmental ethics center on whether justice for all applies only to humans. These debates usually center on the presence of intrinsic value in nature and whether anthropocentrism, biocentrism, or another position is the right theoretical foundation for environmental ethics. Yet these positions lead to the same broad result when combined with the rest of the principles. Namely, adequate assessment indicates whether one is anthropocentric or biocentric; in deciding whether one agrees with the extension of intrinsic value to otherkind or not one will need to consider the needs of humans *and* otherkind when constructing environmental policies because at minimum, human needs cannot be served without responding to the needs of the rest of nature on which we depend for our existence. Thus, one is responsible to and for one's actions with respect to all entities and must carefully consider how one uses natural resources. Thus justice generally refers to a sort of distributive justice wherein entities get what they deserve based on morally relevant characteristics. The definition of such characteristics and the extent to which they are systematically articulated may vary greatly in different ethical systems, but the principle of justice sets a broad bound of ethical consideration which, as we shall see, provides a significant corrective to existing indexes of sustainability.

As justice expands our circles of consideration, *farsightedness* expands the spatial and temporal scales considered in decision-making. Farsightedness, like "Long-range thinking" which is listed in Louke van Wensveen's list of virtues in the appendix of *Dirty Virtues*, implies an eventual goal of internalizing long-range thinking (van Wensveen 2000: 164). But farsightedness intentionally looks at spatial and temporal dimensions. While the sustainability literature has largely focused on intergenerational issues – the far future – separating farsightedness from sustainability highlights temporal and spatial issues to be limited such that the present and future, one's immediate surroundings and distant places, are connected. With such a focus, and coupled with justice and responsibility, farsightedness enables sustainability literature to not only focus on the temporally and spatially distant ramifications of decisions but also to keep current, but often out of sight, problems in view. For instance, those who may be overlooked in typical environmental analyses because of their race, ethnicity, citizenship, economic status, gender, ideology, or species may be considered.

Though farsightedness aims to extend the scope of consideration to all times and places, and categories of concern, one cannot practically consider every moment and every entity from now to eternity in all places when making a decision. Since some periods are more relevant than others with respect to a particular set of circumstances, considering the duration relevant to a decision, and erring on the side of caution in the face of inevitable uncertainties by assuming that new interactions with the environment will be unacceptably harmful unless demonstrated to be otherwise, is a safe approach consistent with the other principles. The goal of farsightedness is to enable people to internalize long-range thinking. At first glance, it may seem that this concept of farsightedness is just sustainability, for sustainability studies typically examines the moral implications of actions in distant time and space. Yet, defining farsightedness separately from sustainability, as one of many ethical principles required, is advantageous because it enables people to recognize the different components of ethics needed for sustainability rather than merging them all into one such that parts are easily forgotten in the shuffle. Farsightedness emphasizes the long-range portion of sustainability while responsibility provides the "ought" that some include in their definition of sustainability. The other principles provide the details by which farsightedness ought to be applied.

Careful use acknowledges that we must use resources to survive but that we should do so with caution. This care may include care for ourselves, others, our relationships, and possibly for resources themselves. Careful use can imply efficiency, recycling, focusing on needs rather than wants, and decreasing the use of the most harmful substances.

Responsibility assumes that humans have caused environmental destruction and ought to do something to rectify the situation and prevent it in the future. Regardless of the theological or philosophical grounds of responsibility, it provides the impetus to act and the motivation to follow the other principles. In this way, it operates at a slightly different level than the rest of the principles but is still deeply linked to all of them.

Feasible idealism aims to ensure that decisions reached through the use of the broad principles are implementable yet do not stray too far from ideals. Unfortunately, there is no one-word description for this principle. Feasibility alone is close, but not enough. If used by itself, feasibility may encourage people to abandon their ideals for the sake of implementing an action plan rather than to creatively think about how to realistically approach their goals. Thus a compound term is necessary to name this principle. Feasible idealism, like adequate assessment of the situation, requires a real interdisciplinary effort. To employ this principle one must know something about the technical, economic, and political feasibility of any proposal including whether the degree of cooperation necessary to achieve the goal is feasible. It is different from adequate assessment, however, because it is consistently forward-looking and involves an imaginative process of determining what is desired in concert with the other principles. Please note that the feasible portion of feasible idealism is not intended to imply that the easiest decision to implement should always be selected. Rather, the political, technical, ethical, and economic feasibility of an option, and its consequences if enacted, must be evaluated. The goal here is to choose actions and policies that have the largest possible positive effects for humans and the environment yet are within the realm of the possible. In this way, people can be stretched to innovate, to meet new environmental challenges, and to improve actions toward and in the environment.

4.3 Potential critiques revisited

Now that the theory of the broad principles has been explicated and a working list of broad principles identified through a process of comparison, the likely questions about this theory raised in the introduction to this chapter, "Why not use the ethics of the Earth Charter?" and "Will the broad principles be acceptable to people with deeply held worldviews including ethical commitments?" can be addressed in more detail.

4.3.1 Why not the Earth Charter?

With these principles in hand, one may wonder how to compare them to the Earth Charter, an ethical document for sustainability produced through a multiyear collaborative process involving numerous people from a variety of worldviews from around the world, especially since space constraints limited the number of ethical systems examined here and the Earth Charter process involved participants from many traditions. Such a comparison finds that the principles of responsibility, careful use, adequate assessment of the situation, justice, and the balance between idealism and feasibility align well with the Earth Charter. The most significant differences arise with the

Charter's attitude toward adaptability, and its specificity with respect to particular values. Let us look at each in turn to better understand the advantages of the broad principles.

The Charter does align with the principle of adaptability in that it acknowledges the need to work at different levels depending on the particular problem at hand. It is willing to adapt policies to uncertain or unknown information; and it tries to resonate with many cultural contexts. However, the Charter does not encourage the revision of the principles themselves as circumstances change. It went through a process of revision to arrive at its current form but there is no provision in place to enable revisions or extensions in the future. While I support the Earth Charter's tenets, I would be more comfortable if it acknowledged that moral knowledge is still developing. At the beginning of the twentieth century, most ethicists would not have predicted the movement toward a global environmental ethic. Because moral knowledge can develop and has developed over time, constructing our ethical theories to acknowledge, account for, and allow this change would more closely align with the process of ethical development and implementation as seen in the world while also enabling responses to complex, dynamic environmental problems.

Finally, the most obvious difference between the Earth Charter and the broad principles is that the Charter's principles are so much more specific: they name the importance of otherkind for reasons other than to fulfill human needs, name the need for family planning to lessen environmental impacts of large populations, and prioritize democracy over other systems of government. They also advocate for certain types of economic changes – that big business and multinationals need to be held accountable for their impacts on societies and the environment. The presence of these details may limit the number of people who are willing to support the Earth Charter (Derr 2000: 12). Whether the specifics of the Earth Charter or the vagueness of the broad principles is more appropriate and more useful for international decision-making depends on whether each set of principles can productively limit the possibility space of moral decision-making and whether one will be able to gain enough support to be influential. It is outside the scope of this project to precisely determine the potential adoption rate of the broad principles as opposed to the Earth Charter or the comparative utility of the two. Rather, I will demonstrate the utility of the broad principles as I use them to evaluate sustainability indexes in the next chapter.

4.3.2 Will the broad principles be acceptable?

Another likely question when encountering the theory of broad principles and the principles themselves is whether people with very firm convictions will be willing to align themselves with the broad principles for purposes of collaborative dialogue and decision-making about sustainability or if they will insist on using their own concepts and terminology, referencing their theological, metaphysical, and axiological beliefs, and refuse to use the broad principles. Certainly it is possible that individuals or groups will refuse to participate in the process of developing and using broad principles. It is also possible that people who are willing to participate in such a process will insist upon referencing their own core beliefs in discussions. Deep ecologists may reference intrinsic value, biocentrism, or the intuitive experience of nature. Muslims may refer to God or Muhammad, adding PBUH (Praise Be Unto Him) to their comments. Similarly, Christians may reference the Trinity, Jesus, or the Bible. These ways of thinking and speaking are prioritized and expected in these worldviews. Such commitments do not, however, mean that adherents of these worldviews will not be able to use the broad principles or participate in such a process. Many ethical and theological conferences and documents of the last century including the Universal Declaration on Human Rights, Agenda 21, the Earth Charter, the Millennium Development Goals, and scores of other initiatives from the local to international level illustrate that such collaboration can happen, at least to some degree. Because the process of identifying the broad principles is not intended to homogenize everyone, but rather to identify commonalities where they exist while preserving the ability of attending to one's deeply held beliefs, presumably many will be willing to participate in this process. After all, the broad principles only provide broad bounds of ethical decision-making and often will need to develop more narrow bounds as decisions are made in particular communities that rely on community-specific specifications of the broad principles. These possibilities, combined with the fact that different metaphysical positions do not necessarily lead one to different actions and the fact that the principles are always open to correction should enable the principles to be utilized by a wide variety of people.

4.4 Implementing the broad principles: a preliminary method

Before the utility of the broad principles can be demonstrated in Chapter 5, questions of implementation should be addressed: How will the principles be put into practice? What would it look like to use the principles? Are there hard and fast rules for their application? Do any principles have priority over others? What happens if the principles conflict? A precise ethical system, one that describes exactly what to do in each circumstance, according to a neat flow chart, may be desirable. After all, such a gilded decision-tree would erase all concerns about doing the wrong thing. Yet no such tidy, certain formula exists or will be developed; real life is too complex and messy. Even simple decisions involve the consideration of facts, moral values, and cultural considerations as well as various time scales and spheres of moral concern, potentially including obligations to self, family, friends, otherkind, religious community, nation, ecosystem, world, and Ultimate. Additionally, as we have seen, the broad principles mutually influence each other and thus need to be considered more or less simultaneously. To further complicate matters, all decisions must be made with incomplete knowledge about the state of the situation and the possible outcomes of one's decision given the complex dynamic world in which we live. But some guidelines are needed to aid the application of these ethical principles, especially since humans are still novices at making informed, collaborative decisions about global environmental issues (Hargrove 1985). Thus, in this section I outline a tentative framework for applying the ethical principles that relies on Eugene Hargrove's work on the role of rules in ethical decision-making. Particularly, I suggest beginning with an assessment of the situation, continuing the ethical analysis with the application of farsightedness, adaptability, careful use, justice, and responsibility individually and in combination, and culminating with a careful assessment of the feasibility of implementing the various options which emerge in the analysis. Admittedly, this is a time-consuming approach, but with time, people who follow it will develop skill at applying the principles so that subsequent ethical analysis will be less arduous.

Hargrove's insightful examination of the role of rules in ethical decision-making is based on an investigation of the use of rules in chess and illuminates a framework for applying the application of ethical principles. Hargrove argues that when people are first learning to play chess or make ethical decisions careful attention to specific rules, whether "control the center of the board" or "never lie," help them understand the implications of their actions and learn how to make good decisions (Hargrove 1985: 10). He maintains, however, that neither experienced chess players nor ethical decision-makers mechanistically follow a rational, universal set of ethical rules and rarely perform elaborate rule-based calculations (Hargrove 1985: 17, 21, 30). Instead, they transcend the use of rules as they develop creative, intuitive approaches to decision-making (Hargrove 1985). Consequently, rules are most helpful in training and justification: when people are first learning how to make ethical decisions and when anyone, novice or expert, explains their ethical decisions to themselves and others (Hargrove 1985: 19). Thus, as long as ethical decision-making with respect to wide-scale environmental destruction involves new elements or decision-makers require justification to the larger community then rules for the application of principles, and the principles themselves, have an important role.

Hargrove argues that people ought to be taught to use ethical rules and allowed to practice so they can learn which principles are most appropriate in particular types of situations (Hargrove 1985: 17). Given the complexities of decision-making about the environment – the massive amount of data involved and the incredible numbers of people that need to have input into major decisions – it will take time to internalize the process. Even when individuals or groups are adept at ethically astute environmental decision-making, they may need guidelines to apply the principles to new scenarios, to ensure that they themselves consider all sides of the issue or justify their positions to others.

In response to these needs, I recommend the following three-part method of applying the principles, though, not surprisingly, I shy away from the word "rules" insofar as it connotes static rather than revisable guidelines. First, the situation must be assessed to determine who and what is involved. The stake various participants have in the situation, how they have been treated in the past, and how they may be affected by various consequences of different options must be identified to the extent possible. To answer these questions, care must be taken to consult the natural and social sciences as well as traditional ecological and community-based knowledge. The ultimate foundations important to the decision-makers such as belief in God or intrinsic value must also be referenced as they may constrain the possible decisions and make their justifications more compelling for those that believe them. Similarly, other ethical principles such as responsibility or justice may shape the assessment of the situation.

Once something of the situation is known, possible options for action should be evaluated according to the other principles. Will some options be better in the far future or distant places even though they may have only mild or no benefits locally or in the short run? Do some options enable and support justice for all more than others? Do some options encourage the careful use of resources? Are some options appropriate for a broader scope of individuals, nations, or biota? Will some be more able to take account of new circumstances in the future? Will some enable people to be more responsible by giving them information that, when coupled with other ethical priorities, may inspire them to act? Will some options encourage responsibility in long spatial and temporal scales by targeting the root of a problem rather than just shifting the problems somewhere else? These and other questions ought to be investigated on their own and in combination with each other.

Finally, decision-makers need to apply feasible idealism to the options they consider. Though balancing idealism and feasibility will certainly be a concern throughout the entire process, it is most helpful after the implications of the other moral principles have been investigated. Focusing on it too soon can limit the consideration of options to those that are known to be feasible, a process that can cause novel solutions to be overlooked. Waiting to employ feasibility allows decision-makers to think creatively as they explore the ramifications of various decisions, identify ideals, and recognize the limitations of current practices. In turn, this information may motivate future research and a more comprehensive approach to environmental action. For example, as we shall see in Chapter 5, sustainability indexes do not currently acknowledge the fact that people of color and the poor disproportionably suffer from environmental damage though promoting justice for all is a key feature of sustainability definitions. A thorough ethical evaluation of the indexes is necessary to identify this and other limitations of the indexes and thus, take steps to overcome them.

Of course, care must be taken not to follow the order of applying these principles so strictly that one misses the implications of connections between the principles or of creative ways they apply to new situations. Given the vast spatial and temporal consequences of many environmental decisions, environmental decision-making using ethical principles would ideally be undertaken by representatives of diverse constituencies to account for the various needs and positions of all entities of moral concern. Yet, as Hargrove points out, applying multiple ethical principles or viewpoints to complex situations and encouraging creativity in moral evaluation will be a frightening process for some people. They will fear that considering diverse moral options will lead to extreme relativism, or that the process will be so complex that nothing will ever get done. Notably, however, people already evaluate different moral options and can learn from experience without falling into extreme relativism. Over time, practice with principles can lead to quicker decision-making that considers the specifics of a situation including the ultimate ideals and overall worldviews of those affected by the decisions as individuals or groups creatively internalize the process (Hargrove 1985: 37–38). Of course, these guidelines for applying ethical principles will still be useful so that expert environmental decision-makers, and those new to the process, are able to converse with one another about the decision-making process.

With these guidelines and the broad principles in hand we can proceed past this chapters' focus on 1) a theory of broad principles which can be specified by different, possibly contradictory specifications yet provide broad bounds for decision-making which enable consensus across worldviews; 2) the identification of broad principles through a rigorous comparison of the ethical systems of Agenda 21, Nash, Llewellyn, and deep ecology; and 3) a basic method for applying these principles in which the broad principles should be applied in a general order beginning with considering the situation, proceeding to employing farsightedness, adaptability, responsibility, careful use, and justice, and concluding with addressing the feasibility of the options under consideration. Now we are ready to demonstrate how the broad principles can guide decision-making when integrated with technical concerns. Chapter 5 is just such a demonstration; it applies the principles to indexes of sustainability and sustainable development.

Note

1 Thanks to Matthew Bower, my research assistant, for suggesting this term during a brainstorming session. Personal Communication August 11, 2011.

5 An ethical examination of sustainability indexes[1]

Chapters 2 through 4 established that technical and ethical aspects of sustainability are intertwined, outlined a theory of index construction, developed a set of broad ethical principles, and articulated a method of applying them. With these results, the heart of this project can begin: identifying the ways in which sustainability indexes do and do not align with sustainability ethics, the aim of this chapter, and, in subsequent chapters, identifying ways to improve the indexes' alignment with ethics. Particularly, this chapter's analysis reveals that the indexes generally align well with responsibility, careful use, and feasible idealism. They follow adaptability in some respects but not others. They fall shortest with respect to adequate data assessment and justice because they rarely track differences in access to environmental benefits and exposure to environmental burdens with regard to subpopulations in a nation or community.

These conclusions begin to fill a significant lacuna in the literature. Although some index developers recognize that normative priorities play a role in index development (Bleicher and Gross 2010: 603, Burger et al. 2010, Olalla-Tárraga 2006, Walter and Stuetzel 2009) and a few even recognize the need to consider diverse ethical perspectives to ensure that the indexes are ethically sound before they are adopted and more difficult to revise (Dahl 1997: 82), this work has not yet been done. Indeed, I know of no ethical analysis of sustainability indexes, other than those which focus on the limits of GDP (e.g. Daly et al. 1989), that examines particular indexes or is more than a page or two in length (Peet and Bossel 2000). The ethical analysis of sustainability indexes in this chapter not only illustrates how such analysis can be done but also suggests directions for future research to ensure that sustainability indexes align with technical and normative elements of sustainability.

To determine whether sustainability indexes align with sustainability ethics, they should be ethically evaluated using ethical principles that resonate with the theoretical foundation of the indexes and the people who will be affected by their implementation. The broad principles outlined in Chapter 4 fulfill these requirements. They align with principles implicit in Agenda 21 and other documents that shaped the definition of sustainability presumed by sustainability indexes. Additionally, as they resonate with ethical systems of multiple religious and philosophical traditions they are likely to be amenable to many people affected by the indexes.

Since hundreds of sustainability and sustainable development indexes exist (International Institute for Sustainable Development 2012, Krank and Wallbaum 2011: 1385) that differ in complexity (from single indicators to indicator sets or multidimensional indexes with dozens of indicators), scale (from a focus on individuals or local communities to regional or national levels), and scope (from examining an aspect of sustainability, e.g. energy sustainability in manufacturing, to sustainability as a comprehensive whole), alignment with existing policies (from ideals as yet not implemented to targets specifically identified in policy), and type of index developer (from professional experts to local lay people) an ethical evaluation of sustainability indexes could fill volumes. Instead of determining the ethical strengths and weaknesses of all indexes, this chapter examines a representative list of indexes to identify patterns of alignment with ethics to critique specific indexes and suggest directions for further index research.

The indexes and indicators chosen for assessment in this chapter include indexes developed for use at national levels including carbon emissions indicators, the three-dimensional index of sustainable energy development, the Wellbeing of Nations, the 2012 Environmental Performance Index, and Eurostat's Sustainable Development Indicators (SDI) as well as a number of indicator sets and indexes for community-based sustainability initiatives. Among the national indexes, three are general indexes of sustainability and two examine a particular segment of the sustainability discussion, energy use. The carbon dioxide emissions index is a single indicator, the Eurostat SDI is a set of indicators, and the other national indexes aggregate multiple indicators as they aim to yield overarching assessments of progress toward or away from sustainability. While many of these indicators and indexes have been constructed as theoretical exercises that may be more or less aligned with policy targets, some, such as the Eurostat SDI, are developed to assess progress toward specific policy goals. Finally, local indexes constructed with considerable input from local stakeholders will be examined to contrast with the national-level indexes constructed by experts. Thus, the ethical assessment of sustainability indexes covers indexes and indicator sets for local to national scales developed by professionals with or without community input.

Because energy use is involved in every dimension of human life and is a significant contributor to environmental destruction, energy indexes can offer a reasonable proxy of overall sustainability yet cover a constrained set of issues enabling a detailed ethical analysis in a relatively short amount of time. Thus, Section 5.1 and 5.2 contain ethical analyses of an energy indicator and index respectively. Sections 5.3–5.6 analyze general sustainability indexes at an international and local level, focusing on the ways in which these indexes have unique ethical features compared to the ethical patterns of the energy indexes.

5.1 Carbon emissions indicators

In the sustainability movement anthropogenic climate change is often seen as the largest sign that humans are living in an unsustainable way. Within climate change studies levels of atmospheric carbon dioxide and rates of carbon dioxide emissions, especially from energy use, receive the most attention because, as of 2004, CO_2 emissions account for 77 percent of anthropogenic greenhouse emissions and because energy use from fossil fuels yields 56 percent of anthropogenic greenhouse emissions in CO_2 equivalents (Intergovernmental Panel on Climate Change 2007: 36). In industrialized nations energy use can contribute even more significantly to greenhouse gas emissions. For instance, according to the US Energy Information Administration (EIA), "Energy-related carbon dioxide emissions account for more than 80 percent of U.S. greenhouse gas emissions" (Energy Information Administration 2009). Thus, it is not surprising that methods of monitoring CO_2 are often used to assess contributions to or activities to slow climate change as well as sustainability itself.

There are a number of ways to monitor CO_2 including atmospheric concentrations; emissions rates for countries, cities, companies, or individual households; and emissions rates per capita or per GDP. As will be shown below, these indicators highlight different aspects of the situation but share strengths and limitations with respect to most of the other ethical principles and the guidelines of index development.

While both emissions and concentrations data can be helpful for environmental decision-making, CO_2 emissions are particularly appropriate measures of the energy sustainability of a company or country. Emissions data is directly related to the activities of the group in question whereas atmospheric CO_2 concentrations reflect past actions of all people since a molecule of CO_2 has an average atmospheric lifespan of around a century and does not stay within city, state, or national boundaries. Thus, CO_2 emissions data enables a fine-grained assessment of responsibility for climate change. When emissions are tracked over time they can also be used to assess the degree to which people take responsibility for their impact on climate and subsequently change their practices.

There are, however, several measures of CO_2 emissions, each of which emphasizes a different aspect of sustainability. The basic method records or estimates the amount of CO_2 emitted by an activity or set of activities, a measure that roughly indicates the total environmental impact of this greenhouse gas (Hecq 1997: 128). Alternatively, monitoring CO_2 emissions per capita facilitates comparisons between countries with significantly different populations. Monitoring CO_2 emitted per GDP aims to indicate the value obtained from each unit of emission, assuming that GDP accurately monitors the value of an activity to a country (Hecq 1997, Tester 2005: 45–47). Tracking the CO_2 emitted per capita or per GDP as well as total emissions can help determine whether changes in emission levels are linked to major social or economic changes (e.g. war or economic depression) or if they are due to changes in the structure of energy use itself (industrialization or a shift from manufacturing to services) which may increase efficiency. While such knowledge is important for evaluating current and potential policies, focusing on the CO_2 emitted per capita or per unit of GDP can divert attention from the environmental impact of emissions since these measures do not express the total amount of emissions. For instance, these measures mask the fact that the total environmental impact of emissions will increase if the population rises more quickly than the CO_2 emitted per capita decreases. Thus, if a single, relatively simple, indicator of sustainable energy is desired, a gross emissions indicator is a better candidate than emissions per capita or GDP because it estimates the potential for emissions to affect the environment.

Emissions indexes are most feasibly influential for policy-making when the index's output value can be easily separable into its component parts, whether sectors of the economy or fuels used. This disaggregated data can highlight either the activities or fuels that most significantly contribute to emissions so that policies can target the most destructive. In the United States, the Energy Information Administration (EIA) enables such disaggregation as it monitors the CO_2 emitted by the country as a whole and by the residential, commercial, industrial, and transportation sectors of the economy. Emissions of each sector are tracked by estimating the direct fuels such as coal and natural gas used by the sector, the electricity purchased by the sector, and the industrial processes attributable to particular sectors (DOE/EIA 2006). In this way, policy-makers can identify sectors that contribute most to climate change, assess whether policies to reduce emissions are working, and determine the areas where efficiency measures or legislation may have the most impact.

While most emissions data concerns gross emissions, some analysts try to monitor net emissions. From a nation's total CO_2 emissions they subtract estimates of emissions offset or sequestered by activities such as planting trees (Hepple and Benson 2005, Herzog and Drake 1996, Holloway 2001, Tester 2005: 46–47, 178–79). Monitoring net emissions improves the correspondence between the index results and the actual rates of climate change but complexifies the monitoring process. Estimating CO_2 uptake for large tracts of vegetation is a difficult process and the technical and economic feasibility of collecting and sequestering carbon is still debated. To avoid this complexity and yield more certain data, mitigation efforts are rarely included in emissions data.

While a narrow focus on gross CO_2 emissions has many advantages, this trend can be a liability if it is the only index used to monitor sustainability as it can enable people to overlook 1) other contributions to climate change, such as methane from agriculture or 2) environmental problems other than climate change. After all, no one indicator can register the complexity of movement toward or away from sustainability. Thus, if a CO_2 emissions

126 *An ethical examination of sustainability indexes*

indicator is to be *the* or a main indicator of sustainable energy development (SED), those relying on the index must constantly be reminded that it is a simplistic picture and that the strategy of monitoring SED may need to be adjusted as other environmental problems grow.

A gross CO_2 emissions index does, however, have a number of ethical advantages. As it highlights the aspects of energy use that will have the largest effect on the environment, human society, and the economy for centuries it is a quite farsighted index. It can help people overcome their local biases to examine the problems most significant in time and space. After all, it is often easier to identify and raise awareness about the environmental impacts of our actions that are visible and local. Water pollution in a city lake or toxic waste near a school typically inspires action more than environmental damage thousands of miles away or years in the future. The general idea of a CO_2 emissions index is also advantageous because it can be adapted for use at a variety of levels from the local to the national with respect to multiple activities or types of fuel used. Given this flexibility, the index can help identify who has been and is less responsible for climate change and who is taking responsibility to limit such effects. Additionally, because the CO_2 emissions index encourages people to reduce their emissions of CO_2 it encourages careful use, though it does not mandate a specification of this principle. People could reduce their CO_2 use through technological advances to increase efficiency, though attitudinal changes that lead to people wanting and using less, through recycling, or a host of other initiatives.

When we examine the index through the lens of justice, we uncover one of its greatest weaknesses: it provides no way to assess the distribution of CO_2 impacts. The global nature of the atmosphere contributes to injustice because the CO_2 released in one area may influence people in far distant areas who do not benefit from the activities that emit it. Developed countries have emitted massive amounts of CO_2 and experience the advantages of high quality energy services. On the other hand, developing nations have emitted relatively small amounts of greenhouse gases per capita. Yet, these countries are and will be most significantly affected by climate change because they lack the economic, political, or material resources to mitigate or move from newly flooded or drought ridden areas, higher temperatures, and so on (Intergovernmental Panel on Climate Change 2007: 48–52, Stern 2006: 59). Since the CO_2 emissions index cannot register the distributional disparity between benefits and harms of CO_2 emissions, it only follows the principle of justice in the most rudimentary way: decreasing CO_2 emissions is an environmental good for people and otherkind as a whole.

Of course, this generalization overlooks the fact that CO_2 emissions may aid biota and ecosystems that thrive in warmer climates or in a CO_2 rich atmosphere. If these entities are prioritized then a different index would be needed to align with adequate assessment of the situation, justice, and responsibility. Generally, however, for the sake of humanity and/or biota in general, environmentalists prioritize past ecosystemic conditions over those that humans are bringing about. These values plus the desire for a simple index yield the directionality and generalization of the CO_2 index.

While the CO_2 index is limited in its scope, as are all narrowly focused indicators of complex systems, we must not lose sight of advantages of this index: its relative simplicity allows it to be more easily understood than complex, multidimensional indexes that track many ramifications of energy use. Additionally, its focus on CO_2 emissions enables it to monitor the largest contributor to climate change, one of the most significant sustainability issues facing the world today. The index also fosters farsightedness, careful use, and responsibility though its ability to foster justice is quite limited at any but the most general of levels. If a more complex understanding of movement toward SED is desired, a more intricate index will be required, though as we shall see, these indexes also have limits with regard to justice.

5.2 The three-dimensional index of sustainable energy development (SED)

The three-dimensional index of sustainable energy development (SED) was developed to capture the complexity of SED better than one-dimensional indexes such as the CO_2 emissions index (Davidsdottir et al. 2007: 305–6). It tracks movements toward or away from economic, environmental, and social dimensions of SED on one three-dimensional graph. This output mechanism allows the index to track simultaneous but different changes in the various dimensions. Thus, it acknowledges that some actions have both positive and negative consequences for sustainability. Its flexibility also enables countries with very different developmental paths to use it and allows policy-makers to assess the efficacy of policies for a dimension and identify aspects of sustainability that most urgently need attention. Ethically, its strengths include the fact that it monitors access to high quality energy services, a key precondition for improvements in quality of life; is committed to long-term sustainability; and is able to foster careful use and responsibility. Its limitations include using proxies in the social dimension due to a lack of data; the difficulty of finding indicators that consistently measure movement toward or away from SED, particularly in the economic dimension; and its inability to monitor inequitable exposure to environmental harms among human communities, a key feature for aligning with justice. Additionally, the energy import indicator is ineffective for monitoring energy sustainability because decreasing energy imports does not necessarily increase the sustainability of energy used.

Given the complexity of the index, Section 5.2 includes five subsections. The first describes the index's goals, components, and calculations. The second, third, and fourth analyze the environmental, social, and economic dimensions of the index using the ethical principles. The fifth subsection ethically evaluates the index as a whole.

128   An ethical examination of sustainability indexes

5.2.1 Foundations of the three-dimensional index of SED

The developers of the three-dimensional index relied upon definitions of and goals for sustainability and sustainable development articulated by major international organizations in order to base their index on as influential, well-accepted, and scientifically sound ideas as possible. Particularly, they followed the International Atomic Energy Agency (IAEA) and the International Energy Agency's (IEA) definition of energy sustainability: "the provision of adequate energy services at affordable cost in a secure and environmentally benign manner, in conformity with social and economic development needs" (IAEA/IEA 2001: 1). To describe the relationships between the dimensions of sustainability and energy use, they relied upon the conclusions of the 1972 Stockholm Conference on the Human Environment, the 1992 Earth Summit in Rio de Janeiro, and the 2002 World Summit on Sustainable Development in Johannesburg, which articulate the relationship between energy and environment, economy, and society (Davidsdottir et al. 2007: 305–6). The Johannesburg definition of sustainable development regarding energy was particularly influential on the index:

> access to reliable, affordable, economically viable, socially acceptable and environmentally sound energy services and resources, taking into account national specificities and circumstances through various means such as enhanced rural electrification and decentralized energy systems, increased use of renewable energy, cleaner liquid and gaseous fuels and enhanced energy efficiency.
> (As quoted in Najam and Cleveland 2003: 133)

To move from this broad definition to the specific indicators comprising the index, Davidsdottir et al. followed the IAEA and the IEA's work to identify indicators appropriate for a composite index of energy systems, "the most comprehensive effort towards identifying SED indicators" (Davidsdottir et al. 2007: 305).

Combining these perspectives, they identified "four central goals ... of SED" (Davidsdottir et al. 2007: 306): First, increasing "the technical and economic efficiency of energy use and production" helps ensure that the negative impacts of energy use including harm to the environment, human health, and societal structures are minimized while promoting the advantages of energy use. Similarly, using energy in economically efficient ways prevents spending on energy from dominating the market and crippling commercial advancement. Increasing efficiency also ensures extended access to energy services as supplies of fossil fuels diminish, a way to improve energy security, their second goal. A secure energy future also depends on diversifying energy carriers, distributing fuel sources geographically, and avoiding relying too heavily on energy imported from unstable regions. Switching to cleaner fuels and energy technologies is a third goal of SED since energy use currently produces more wastes than ecosystems can process, conditions that do and will harm the environment, human health, and the economy. The final goal is a type of justice: Davidsdottir et al. note that even if energy services are efficient, secure, and environmentally sound they will not fulfill the popular dream of sustainability unless all people have affordable, sufficient access to energy services to meet their basic needs. Indeed, all of these goals aim to ensure that energy use is possible and sustainable for people today and in the far future with minimal environmental disruption (Davidsdottir et al. 2007: 306).

In the articulation of these goals, Davidsdottir et al. generally emphasize humanity over other aspects of the environment. For instance, they clearly want to sustain human society and well-being as they strive for a strong economy, peace, and justice among humans, though they also prioritize staying within the carrying capacity of the environment and decreasing harm to the environment. It is unclear whether their environmental dimension is included primarily or only because of its importance for human life, or if it is also valued for its own sake. Because their rhetoric emphasizes humanity it is likely that it will be critiqued by at least some proponents of intrinsic value who would desire a greater emphasis on environmental conditions and biota to adequately assess the situation from their point of view and promote responsibility and justice for more entities than humans. The index, however, does not have to be interpreted as only valuing humanity. After all, its environmental indicators track pollutants that have an impact on the well-being of otherkind as well as humanity. Even its social and environmental indicators are environmentally significant because poor societal and economic conditions can drive further environmental degradation. Adjudicating whether or not the index is problematically anthropocentric, and if so, how to improve it, requires a particular specification of the broad principles in a particular environment and thus, is beyond the scope of this project.

Aside from possible questions about anthropocentrism, the overarching goals of SED listed above will resonate with the sustainable energy development goals of many if not all countries yet nations will instantiate these goals in different ways. Davidsdottir et al. maintain that there is no final, ideal state of SED because the interaction between the dimensions is complex and always evolving. Additionally, countries take different paths toward SED because their environments, economies, and governments are in different states. Countries that have already achieved affordable access to high quality energy services for most people will need a different plan to move toward SED than those who need dramatic improvements in access while also making progress in other dimensions. Countries which prioritize affordability and access to high quality energy services in the short term will take a different sustainability path than those who opt for a more simultaneous improvement of the dimensions. The three-dimensional index is intended to be flexible enough to take account of both the starting points of individual countries and their various paths toward SED (Davidsdottir et al. 2007: 307–9).

To achieve this state of adaptability, the three-dimensional index measures the simultaneous movement toward or away from SED in the social, economic, and environmental dimensions, each represented on an axis of a three-dimensional graph. Ideally, nations would continually move toward SED, or at least not move away from SED, in all dimensions at all times. Sometimes, however, it may be necessary to compromise one dimension for gains in another, especially in the short term. The three-dimensional graph enables the index users to recognize these tradeoffs and thus, the complexities of various paths toward sustainability.

Each dimension of the index is comprised of two to four indicators selected from the IAEA/IEA list of forty-one indicators for SED (IAEA/IEA 2001: 7). Indicators were included in the index if they could be calculated with data readily available from most countries and if their inclusion did not cause redundancy in the overall index. Both Driving-Force and State indicators were used to encompass a larger picture of a country's progress toward SED.

Once data is collected the indicators are computed for each year data is available. Each indicator's movement toward or away from SED is calculated by its change from the prior year divided by its value during the prior year. The value of one dimension of the index is set equal to the average change of its indicators from the previous to current year (Davidsdottir et al. 2007: 317–18). Repeating the process for each dimension yields a three-dimensional point for each year in which data is available for a country. Graphing the points on a three-dimensional graph generates an overall picture of a country's progress toward SED over time. Differentiating the index results by dimension illustrates the improvements of or decline in each dimension.

5.2.2 The environmental dimension

The environmental dimension of the three-dimensional index seeks to monitor the environmental impact of energy production and consumption by humans. To balance a thorough assessment of the complexities of energy use with the need for simple, transparent, and easy-to-use indexes, it is comprised of four indicators representing different scales of environmental effects of energy use. Local, regional, and global pollution are represented by the emissions of nitrous oxides, sulfur dioxide, and carbon dioxide respectively. Admittedly, these emissions can have effects at spatial scales other than those they represent in the index. For example, NO_2 can contribute to global warming. However, each of these pollutants is most influential at the scale they represent and as such serve as decent indicators for environmental damage at the various scales. The fourth indicator focuses on the temporal scale of environmental disruption and a different type of risk altogether: that from accumulated, nontreated, spent uranium. Decreases in the value of these indicators is considered movement toward SED; declines in their averaged sum indicates movement toward SED by the environmental sub-index as a whole (Davidsdottir et al. 2007: 315). With this overview of the environmental dimension in hand, the ethical assessment of these indicators can proceed.

The environmental indicators were chosen because they are some of the most significant pollutants relating to human energy use, because data regarding them is relatively widely available, and because they represent different scales of environmental impacts, three signs that these indicators aim for an adequate assessment of the situation. Certainly, more indicators monitoring methane emissions, runoff, ground water contamination, and changes to the landscape from mining, drilling, transporting, and refining fossil fuels as well as the wars and exploitation of people which can result from obtaining such resources could be included. Davidsdottir et al., however, aimed to keep the index relatively simple so they focused on indicators with standardized methods of measurement and reliable data. This decision also emphasizes indicators most directly linked to energy use. For example, approximately 40 percent of methane emissions result from fossil fuel industries or biomass burning (Houghton 2004: 43, 253) so including it in the index would move the index focus away from energy sustainability unless only energy related methane emissions were tracked, which would add significant complexity to the index. Similarly, environmental impacts such as biodiversity loss or species extinction are not included in the index for while they may be partially caused by energy use, determining the extent to which energy use contributes to such results is too involved for the index. Though there are reasons for limiting the indicators in the environmental dimension, the principle of adequate assessment of the situation does encourage research into additional environmental impacts of energy use to ensure that significant energy-related environmental disruption is not overlooked, a sentiment endorsed by the index developers themselves (Davidsdottir et al. 2007: 315–16).

As we continue to evaluate the environmental subindex with the principles of adequate assessment, observe that the index uses data regarding emissions or accumulated pollutants rather than concentrations in order to correlate the activities of a particular country with environmental degradation. Yet the application of adequate assessment of the situation reveals wider implications of the indicators. For instance, consider the nuclear waste indicator: it monitors the amount of accumulated, untreated spent fuel from the nuclear power industry. If the index is applied to nations in isolation from each other, then shipping nuclear waste overseas would be registered as an improvement in the sustainability of the country generating the waste and a movement away from sustainability in the country accepting the waste. If the index is ever used to assess the efficacy of policies then policy-makers and index developers must work together to decide whether this use of the indicator is acceptable. After all, the country's energy use is not more sustainable if waste is shipped elsewhere; the waste still exists and could still harm people and ecosystems whether in or outside of the country that generated it given the interconnectedness of ecosystems and the long lifetime of nuclear waste.

The problem of country boundaries is not, however, limited to questions of nuclear waste disposal. Many social and economic externalities may be generated, stored, or experienced in one country for the benefit of another (Bastianoni et al. 2004: 254–56). For example, the emissions levels recorded in the index do not include emissions produced by manufacturing goods overseas for the benefit of the country whose index is calculated, neither do they exclude emissions produced when manufacturing goods for export. Theoretically, one could track the amount of energy used to manufacture each product, the types of energy used, and the emissions rates of each fuel to monitor such exchanges but doing so would require more detailed and extensive data than is currently available or that would be feasible to include in an index. Thus, applying the principle of adequate assessment of the situation yields suggestions for further research about indicators, but is not able to fully suggest replacements given the state of the literature and the need for feasibility.

The ethical assessment of the environmental dimension of the index is not, of course, limited to adequate assessment of the situation. Applying the principle of adaptability also identifies strengths and weaknesses of the index. The index cannot track variations in pollution rates within a nation and therefore cannot register patterns of environmental risk to various human communities within a nation. Even if the indicators of the environmental dimension and the value of the environmental subindex as a whole indicate that environmental conditions are improving, there may be, as there often is, a disproportionate impact of environment destruction on people of color and the poor (Agyeman et al. 2003, Bullard 1994c, Harlan et al. 2008, Intergovernmental Panel on Climate Change 2007, Stern 2006). To monitor such changes, indexes would need to be able to adapt to monitor the environmental burdens and benefits of subpopulations of a nation. Since this is not possible with current data, policy-makers must have another way of assessing progress toward environmental justice or, at minimum, recognize the limits of the indexes if they want their work to align with the vast majority of sustainability discussions which prioritize justice.

The environmental dimension pertains to industrialized nations more than developing nations. After all, it focuses on fossil fuel pollution and nuclear waste, problems most pertinent to industrialized and industrializing nations. This emphasis is reinforced by the fact that industrialized nations often have financial resources and expertise necessary to collect the nation's data to employ the index which developing nations may not have. Indeed, as a country switches from nonmarketed biomass fuel to marketed fossil fuels it is easier to collect energy statistics because of the record keeping involved in formal markets. For the index to be applicable to the situations of all countries, it ought to consider more pollutants, particularly particulate matter from burning biomass indoors since respiratory illnesses caused by indoor biomass burning are the fourth largest global health risk, contributing to approximately 1.6 million deaths per year (UNDP 2005: 8). Certainly, the index developers intend to include more environmental indicators when appropriate data are available. Until then, it is important to remember that the index does focus on the environmental problems associated with the most spatially and temporally significant environmental effects of energy use and is most applicable to developed countries.

Though the index emphasizes the impacts of energy use within developed countries, it is adaptable to various conditions and priorities within countries insofar as it allows movement toward and away from sustainability to be simultaneously measured in different dimensions. Countries that currently are emphasizing improving any one dimension of the index can utilize it as can countries which aim for the simultaneous maximization of all of the dimensions of the index.

As the index tracks rates of emissions and accumulated nuclear waste it lays a foundation for responsible action as it enables people to track the results of their decisions over time, assuming that levels of known pollutants correlate with higher environmental risk. At the same time, it is farsighted as it forces people to recognize that even though some pollution might be deemed an acceptable risk in the short run to quickly improve human well-being, it will likely cause movement away from environmental sustainability in the short and long run.

While the environmental dimension is aligned, to some degree, with the principles of adequate assessment of the situation, adaptability, farsightedness, justice, and responsibility, it most fully and most obviously exemplifies the principle of careful use. Here, either energy resources or the regenerative capacity of the environment can be carefully used. Reducing the use of fossil fuels and nuclear materials outright and therefore reducing the pollutants associated with their use could occur through lifestyle choices that decrease energy use or through efficiency measures if the number of people using energy does not rise faster than reductions in energy use and if efficiency measures do not encourage people to spend their money on additional energy-consuming products. Such endeavors would ensure that energy resources were available for future generations while reducing contemporary environmental impacts. Emissions and the accumulation of waste could also be reduced through scrubbing emissions from power plants, recycling nuclear fuel, offsetting emissions, or using cleaner fuel sources, methods of ensuring that the regenerative capacity of the environment is used carefully. All of these methods move the environmental indicators toward SED in ways that align with careful use.

Yet the discussion of the indicators, and the documents on which they are based, rarely mention, let alone emphasize, the reduction of consumption through lifestyle changes. Thus, the many advocates of the frugality specification of careful use will probably argue that while the indexes themselves are appropriate the rhetoric surrounding them could and should be expanded to encourage simpler, less energy-consuming lifestyles. In any case, the index itself, if used, will promote careful use in its broad sense.

134 *An ethical examination of sustainability indexes*

In sum, the environmental dimension of the three-dimensional index has significant strengths and some weaknesses when assessed according to the broad ethical principles. The dimension is well-aligned with the principles of careful use and farsightedness. With respect to adaptability, responsibility, and a realistic assessment of the world, it does fairly well though it is currently hindered most by data limitations. Because the environmental dimension neglects to study the disproportionate effect of environmental destruction within and between countries it falls short of embodying justice. Despite these limitations, the environmental dimension is an ethically significant portion of the index because it prioritizes the most significant, far-ranging and comprehensive measures of energy-related environmental destruction while focusing on the available data. Similar results are found when analyzing the social dimension.

5.2.3 The social dimension

To distinguish itself from the environmental and economic dimensions of sustainability, the social dimension should monitor the well-being of individuals and groups and/or the state of cultural resources. The social dimension of the three-dimensional SED focuses on the well-being of individuals because this parameter is seen as a precursor to societal well-being and individuals themselves are valued. Recognizing the correspondence between the use of high quality energy services and increased quality of life for individuals as registered by the Human Development Index (HDI), the social dimension of the three-dimensional index of sustainable energy development monitors whether people have access to high quality energy services and can afford such services. Access and affordability are separated into two indicators because each is necessary but insufficient for an individual to benefit from high quality energy (Davidsdottir et al. 2007: 311–12). When evaluated ethically, the social dimension aligns well with the broad principles given the available data but if it called greater attention to the limits of data it could better follow justice, careful use, and adaptability to the situation of developing countries which in turn could prompt more responsibility and farsightedness. To arrive at these results, the affordability and access indicators are described and then evaluated ethically.

In the three-dimensional index of sustainable energy development affordability is assessed by the ratio of private expenditures on fuel and electricity per capita to the disposable income per capita. This figure yields a rough approximation of whether high quality fuels are affordable for the average person; lower ratios indicate greater affordability.

Ideally, the affordability of energy would be assessed by the ratio of the average private expenditures on fuel and electricity of the poorest people in the country to the average disposable income of the same group because the affordability of high quality energy for a person with the nation's average income does not necessarily reveal anything about the affordability of energy for the poor. Conversely, if high quality energy is affordable to the poorest people it will also be affordable for those with average and high levels of financial resources. Thus, monitoring the affordability of high quality energy for the poorest people, say the poorest 20 percent of the population, would indicate the affordability of energy for all. Unfortunately, international data about the income of the poorest people is not available, so data for the average population is used (Davidsdottir et al. 2007: 312–13).

Yet affordability is not a sufficient measure of the social dimension of energy sustainability because people would spend less on energy if energy services were unavailable or disrupted to the point that fuels could not be bought. A measure of access to high quality energy is also needed. Davidsdottir et al. note that they would have ideally monitored access through the fraction of households heavily dependent on noncommercial energy without access to electricity because this figure represents the share of people who do not have access to the highest quality energy service, electricity (Davidsdottir et al. 2007: 312–13). Decreasing this fraction would suggest a movement toward sustainability. Since reliable data about household dependence on nonmarketed energy is not internationally available, they used electricity consumption per capita as a proxy for access. They assumed that a larger percentage of the population will have access to electricity in countries where the electricity used per capita is higher. Up to 2000 kWh/year per capita is registered as movement toward sustainability because the major international measure of quality of life, the Human Development Index (HDI) and its individual components (educational measures, life expectancy at birth, gross national income per capita), generally increase substantially with increased energy use up to this threshold. After this mark, the HDI does not increase significantly, regardless of electricity consumption. Consequently, increased electricity use beyond this mark is taken to be unsustainable in the index because it confers fewer and fewer advantages on society and yet continues to have negative effects (Ahlen 2004, Reddy 2002: 117–19).

With this understanding of the social indicators of the SED let us move to their ethical analysis, beginning with the assessment of the situation and the access indicator. The 2000 kWh threshold beyond which energy use is taken to be unsustainable does have a number of ethical advantages. First, according to many religious-moral traditions from around the world increased consumption of material objects beyond a certain threshold does not guarantee increased happiness or good. Second, many legal and ethical reflections about sustainability argue that resources ought to be conserved to ensure the ability to use them in the future. Third, some people claim that entities often perceived as mere resources should be used carefully because they have value in and of themselves. Finally, it is recognized that resource use can push beyond the waste-assimilative capacities of the environment, causing significant environmental degradation and negative social impacts. For all of these reasons, increases in electricity consumption are not necessarily movements toward sustainability. Indeed, at some point the costs of increased electricity use will outweigh the benefits. Examining the literature, it seemed that after around 2000 kWh per capita per year, increasing electricity consumption yields significantly diminished returns as measured by the HDI and its individual components (Reddy 2002: 117–19, Smil 2003: 97–105).

Davidsdottir et al. note that the 2000 kWh per capita per year threshold of electricity use may very well be technically unsustainable given the number of people using electricity and the way in which the electricity is generated. The environmental and economic dimensions of the index are designed to monitor these aspects of energy use. To avoid double counting, and because it is impossible to determine an exact amount of electricity or energy whose consumption is sustainable, the 2000 kWh per capita mark is used as a turning point of social sustainability in the index (Davidsdottir et al. 2007: 313).

Yet this cutoff does have limitations. First, it is assumed that the quality of life attainable at 2000 kWh given today's energy generation and distribution systems are morally and physically adequate. Second, the index developers presume that no important measures of human well-being are only attainable with higher quantities of energy use. Many aspects of human life are highly valued in addition to the HDI's components including, but not limited to, freedom, participation in political processes, having supportive relationships, peace, security, environmental safety, and experiences of the natural world. So far, such measures of well-being are not standardized, not tracked internationally, and have not been correlated with energy use. If ethical priorities demand that higher levels of quality of life are necessary and technical studies show that they require more energy, then the threshold level may need to be increased.

Conversely, if improvements in energy efficiency enable a similar quality of life with less energy use, the threshold level of 2000 kWh per capita should be decreased. Additionally, current measures of electricity use per capita do not take offgrid wind or solar power into account. If these technologies begin to provide a significant amount of electric power then either they should be counted in statistics of per capita electricity use or the turning point of 2000 kWh per capita of ongrid electricity should be reduced accordingly.

Finally, while electricity use may be more neatly correlated with increasing HDI levels than total energy consumption, electricity is not the only energy source that can improve quality of life. Indeed, many people use multiple fuels including natural gas on a daily basis; ignoring their ability to increase quality of life is problematic. Further studies are needed to understand the relationship between quality of life and high quality nonelectric energy services to determine whether the index is appropriately or inappropriately skewed toward electricity consumption. Until this additional research is completed, focusing on electricity helpfully aligns the index with the most significant known contributors to SED. After all, electricity has many advantages as an energy carrier over the direct use of fossil fuels including the reduction of indoor air pollution and onsite fuel storage.

Access to electricity is one of the more adaptable indicators in the index because the threshold built into this indicator acknowledges that different consumption rates indicate different movements in relationship to sustainability. Additionally, the presence of the cutoff itself indicates the openness of Davidsdottir et al. to adaptation since the cutoff was not in early versions of the index, which registered all increases in electricity use as moves toward sustainability. Presumably, Davidsdottir et al. would be open to more adaptations of the index in the future. Likely candidates for adaptation include registering other forms of energy besides electricity and considering the goals of social sustainability in more detail in order to be applicable to various types of countries. People in developed countries have almost complete access to high quality energy. Thus, index developers and users will need to continue to think about whether developed countries have already reached social sustainability with respect to energy use or whether there is room to improve, and thus a need for new indicators. To think about these questions which relate to the long-term aims of social sustainability would also take a step toward improving the farsightedness of the social dimension. Aside from these questions, the access indicator is already somewhat farsighted since it recognizes that increasing consumption infinitely, or even beyond certain finite levels, is unsustainable.

Insofar as the indicator monitors electricity as a way to assess access to basic needs it aligns with the principles of responsibility and justice. Knowing when people do not have access to high quality energy services can help prompt people to acknowledge their complacency in the situation and revise policies to encourage just patterns of access to energy. Yet the indicator's alignment with justice could be deepened. Focusing on whether the poorest people have affordable access to energy would more directly assess whether policies and activities are aiding all people than figures based on average consumption and affordability since extravagant consumption among a few may mask large segments of a population that have little or no access if average data is used.

The social dimension as a whole could also significantly promote justice, and subsequently responsibility, by registering the distribution of the negative effects of energy use. People of color and the poor are disproportionately affected by air, water, and soil pollution as well as aesthetic and cultural impacts from fossil fuel extraction and refinement (O'Rourke and Connolly 2003, Shrader-Frechette 2002). They are also less likely to benefit from new environmental legislation as grandfather clauses often exempt the most egregious polluters from new regulations (Gorovitz Robertson 2008: 537–39). They may also be less likely to benefit from swift and severe enforcement of environmental protection laws and have fewer resources to respond to environmental degradation by moving away, obtaining good healthcare, or purchasing safe food and water. While the idea of environmental justice has become more widespread in recent years, the concept itself and attempts to measure it have not yet penetrated the sustainable index literature. Without this information it may seem reasonable to assume, especially when in search of a feasible indicator, that progress made toward sustainability will benefit all people. Index developers appear to assume just that. Yet when one knows how often environmental benefits and burdens are disproportionately distributed through populations, it seems unlikely that general movement toward environmental sustainability will significantly benefit all people unless specific attention is paid to this issue when constructing environmental laws. Admittedly, the three-dimensional index of sustainable energy development did not include such assessments because a) the IAEA/IEA set of indicators from which the index was constructed did not include a distributional measure, and b) a thorough standardized measure of the distribution of environmental degradation among human populations has not been developed and the three-dimensional index aimed to be implementable. At minimum, however, index developers could note the desire to include such measures to begin to call attention to the problem and encourage additional research on the subject.

The affordability indicator's reliance on a proxy (the ratio of the average amount of disposable income spent on fuel to the average disposable income) rather than the portion of the poorest 20 percent of the population's income spent on fuel also limits its alignment with justice. Potentially vast disparities between the rich and poor are unnoticed by the indicator when the average fuel affordability is monitored. If meeting the needs of all people is the goal, as is stated in numerous definitions of sustainability, then focusing on the poorest rather than the average will ensure that the needs of all are being met. Until sufficient data to do this becomes available, studying the average population can shed some light on the subject. If high quality energy is not affordable for the average population it will not be affordable for the poorest. Therefore, the proxy can be helpful to assess progress toward social sustainability, especially in countries where large segments of the population cannot afford high quality energy services, even though it would align better with justice to focus on the affordability of high quality energy for the poor.

Another critique of the social indicators arises from the combination of realistic assessment of the situation, farsightedness, and adaptability. Biomass is a common fuel but since it is usually not marketed it is not counted in many national statistics of energy use. Thus, when people transition from burning twigs, grasses, and dung that they or their children gathered to marketed fuel sources the percentage of their income spent on fuel will rise; any percent is more than none. Thus, the social dimension will register a move *away* from sustainability even though the transition may have many social advantages. For instance, if women and children do not have to spend up to five hours a day searching for fuel, they will have more time to work for profit, go to school, or pursue other activities. Affordable access to high quality energy services also decreases the air pollution in homes and respiratory illnesses associated with them (UNDP 2005: 4–5, 12). Thus, less of a household's total assets including time, health, money, and so on may be spent on fuel after a switch to marketed, high quality fuel even if the percent of their money spent on fuel increases. If there was a way to monitor total assets spent on energy, or even time as well as money, the indicator would be more applicable to countries experiencing this energy transition as well as developed countries. Not surprisingly, this data is not available. Recognizing its importance and calling for its inclusion in the index, especially if it was widely adopted, could push index development toward a greater adaptability, farsightedness (the index would be applicable to more countries for a longer period of time), and a better assessment of the situation.

The affordability indicator also aims to promote justice and responsibility by determining and publicizing the affordability of fuel. It presumes that knowledge coupled with ethical values can motivate people to act responsibly to ensure that fuel is affordable to all. Yet when we apply farsightedness to this indicator, we see that increased energy use, with all its related environmental problems, is likely if energy becomes more affordable. Consequently, it is essential to pair the affordability indicator with those in other dimensions to recognize the environmental and economic impacts of immediate, narrowly defined social gains.

In sum, the social dimension aligns with the broad ethical principles fairly well. The dimension's strengths include its ability to register movement toward and away from sustainability depending on the amount of energy used, look to the far future as it measures affordability, foster responsibility by calling attention to social aspects of energy use, and emphasize a very basic measure of social justice through the choice of its indicators. Of course, focusing on the poorest portion of the population and measuring the distribution of the negative effects of energy use would significantly improve the index's alignment with justice, responsibility, and adaptability, as would measures of biomass fuels. Until more data is available, however, these two social indicators roughly fulfill the goals of this dimension as they provide a general estimate of the affordability and accessibility of high quality energy services. Unfortunately, the economic indicators are less adequate.

5.2.4 The economic dimension

Acknowledging the links between the economy and energy sustainability, the three-dimensional index considers the goals of the economic dimension to be 1) improving energy security, 2) increasing in the efficiency of the generation and use of energy, and 3) ensuring continued economic development. Three indicators aim to accomplish these goals: the energy intensity of economic activity, the share of the total primary energy supply that comes from renewables, and net energy import dependence. Evaluating these indicators according to the broad principles reveals that the energy intensity indicator is a decent proxy for energy efficiency as long as it is paired with a measure of economic strength, a measure of renewable energy, and the other dimensions.

Additionally, the renewables indicator follows the broad principles fairly well though limited data hampers its ability to align with adaptability and justice. Finally, the energy import indicator does not sufficiently align with the realities of energy importing or the goals of the index and so should be removed from the index.

One of the main goals of the economic dimension of sustainable energy development is to "increase the end use and supply efficiency of the energy system" (Davidsdottir et al. 2007: 309). It is thought that doing so will confer multiple advantages on the economy including expanding its productive capacity and generating wealth. Assuming constant energy costs, if less energy is used to produce the same level of output, more money will be available for research and development, investing, and improving quality of life. Such conditions are considered beneficial for economies, provided that they are not largely dependent on short-term sales of energy.

Unfortunately, constructing an indicator to measure the energy efficiency of a country's activities is far from straightforward. Possible measures range from physical indicators based on the material use of energy that are notoriously difficult to aggregate given the physical variations in fuels to purely monetary measures that are as dependent on the price of energy as on the material efficiency of fuels.

Between these extremes are efficiency measures that can be aggregated easily yet register the physical amount of energy used. One measure that fills these requirements is calculated by dividing the energy used (in BTUs) by the monetary value of the output of the energy use. Since output is measured using dollars the intensities of different activities can easily be aggregated. At the same time, the reliance on thermodynamic measure of energy use rather than its cost entails that the indicator directly links the amount of energy used to the services it enables. Multiple studies recommend using the ratio of Total Primary Energy Supply (TPES) to GDP as a measure of the energy intensity of the economy in indexes of SED (IAEA/IEA 2001: 7). TPES is the total amount of energy used in the residential, commercial, industrial, transportation, and electric utility sectors and GDP measures the monetary value of the final output exchanged in the market.[2] Since both TPES and GDP are a part of standard data sets for many countries around the world, this index is easily calculated. Given its popularity and readily available data, the developers of the three-dimensional index included the energy to GDP ratio as an inverse indicator of SED; declining energy use per GDP indicates a movement toward SED (Davidsdottir et al. 2007: 309).

Unfortunately, when the value of goods and services is tracked using monetary measures it is possible that the indicator falls for reasons other than increasing energy intensity of the economy. For instance, the ratio could decrease because the price of services increases rather than because the amount of energy used decreases. To avoid this problem, Davidsdottir et al. adjusted the yearly GDP according to the rate of inflation relative to a base year. Yet the ratio is also limited because sectorial changes in the economy including the introduction of new products and changes in imports and exports may change GDP and TPES though the nation may still be responsible for the same, or more, energy use. The three-dimensional index cannot register such trends nor the fact that decreasing energy to GDP ratios of developing countries are primarily due to transitions to high quality fuel sources (Stern and Cleveland 2004).

On a larger scale, the intensity indicator does not quite match the goals of the economic dimension. Remember that the goals of the economic dimension are to monitor energy efficiency and security while ensuring that the economy of a country as a whole is not crippled by sustainability efforts (Davidsdottir et al. 2007: 306). The intensity indicator can register whether the energy intensity of the economy is decreasing, but cannot measure whether the economy is strengthened, weakened, or unchanged. Consequently, it should be paired with a measure of economic health to meet the goals of the index.

Two broad solutions to this problem are available. First, an additional indicator of the economy's strength could be added to the economic dimension. A green accounting measure would be more appropriate than GDP given the many known limits to GDP (Daly et al. 1989). If this approach is taken, the choice of indicator should be made by balancing the practicality of data collection and the desire for comprehensiveness, a balance necessary in all index construction.

Alternatively, one could rely on the interdependence of the index's three dimensions to resolve this problem. If it is assumed that a strong economy is ultimately desired to ensure the well-being of society rather than as an end in and of itself then the social dimension's indicators can serve as a proxy for economic strength. Presumably, an economy would need to be relatively stable for its weakest members to have affordable access to high quality energy services. Consequently, the two social subindicators (affordability and access to electricity) may effectively monitor economic security. Certainly, such a measure differs from the common measures of economic strength such as GDP. Empirical studies are needed to determine the correlations between the conditions necessary to allow affordable access to high quality energy services and a strong economy as tracked by traditional measures and expert assessments to validate this approach. Assuming that such links can be demonstrated, this approach is more advantageous than adding a separate indicator of economic strength since it keeps the index simple and highlights the well-being of people and the environment, rather than the economy for its own sake. Thus, the index as a whole already includes a method of monitoring the strength of the economy, so the energy to GDP ratio monitors energy security. If the ratio is falling and the social dimension indicates the economy is strong, then the country in question is moving toward economic SED.

So far, this study of the energy intensity indicator has focused on the perspective of the principle of adequate assessment of the situation, yet other ethical principles are also pertinent to the subindicator. The three-dimensional index developers aimed for the index to equally and justly apply to both developing and developed nations. The intensity index, when coupled with the other dimensions, meets this criterion. Developed countries generally already have a high quality of life and need to focus their efforts on increasing energy efficiency and decreasing pollution while maintaining their well-being, desires registered by the index. On the other hand, developing countries probably need to increase their energy to GDP ratio as they improve their well-being, standard of living, and environmental conditions but doing so will likely strain the economy, at least temporarily and so the economic dimension registers an increasing energy to GDP ratio as a movement *away* from SED. While it may seem unfair to classify increasing the energy intensity of the economy as a movement away from SED for all countries since such increases are all but necessary for some developing nations if they are to increase their quality of life, remember that the social dimensions of the index tracks the changes in conditions which impact quality of life, so the three-dimensional index as a whole will register net movement to sustainability if this occurs. Individual countries will need to determine how they choose to move toward SED. Knowing that policy changes intended to be beneficial in one dimension may set back another dimension is a part of understanding the complex paths toward SED. All too often environmental damage has continued because people prioritized one dimension above all others. Acknowledging that increasing the energy intensity of the economy may both help and harm developing nations is truthful to the situation and can promote taking responsibility for the effects of one's actions in the past, present, and future.

While the energy intensity of the economy indicator promotes responsibility on this large scale, some of its implications are questionable from a standpoint of responsibility and farsightedness. For instance, moving manufacturing overseas could lower the energy intensity of a country as measured by the index because heavy manufacturing uses significant amounts of energy. Yet, the country in question will still use the goods manufactured with energy-intensive procedures and thus will indirectly use the energy required to make the goods. Consequently, the total energy used for the benefit of a nation may be much higher or lower than is registered by the index. Limiting knowledge in this way makes it difficult for nations to take responsibility for their actions because they do not know the scope of their energy use. Tracking the embodied energy of goods and services used in a country, even if produced elsewhere, could help provide the knowledge necessary to spark and enable responsible decision-making. At the moment, such studies cannot be included in the index: highly aggregated studies are not detailed enough to track all of the trades between countries and embodied energy studies of particular products are time-consuming and scarce. For now, those who use the index must remember its limitations to keep its implications in perspective.

Although the intensity indicator may not take account of all energy used for a nation, it can promote careful use by reducing energy use per economic unit through increasing efficiency in manufacturing, recycling, and decreasing consumption of the most energy-intensive products and services. All of these changes are registered as movements toward sustainability by the indicator. Yet maximizing the indicator does not necessarily entail using less energy. As the economy grows, its rate of growth could outweigh increases in energy intensity such that more energy is used. Careful use suggests that such increasing energy use is problematic because it causes negative environmental and social impacts in the present and can decrease the amount of fuels available for the future. Additionally, if the resources are considered a good in and of themselves, they ought not to be squandered. Thus, the intensity indicator must be paired with the indicators of the environmental and social dimensions if it is to align with careful use.

To review, the energy intensity indicator is a decent proxy for energy efficiency as long as it is paired with a measure of economic strength and indicators from the other dimensions. As the indicator is used, its limitations, whether resulting from the urge to have simplistic indicators, or from the incompleteness of data sets, need to be remembered to ensure that the principles of careful use, adequate assessment of the situation, and responsibility are upheld. Regardless of these limitations, the indicator is strong with respect to the principles of adaptability, farsightedness, justice, and general careful use, especially when paired with the other indicators and dimensional subindexes in the index.

The second major goal of the economic dimension of the three-dimensional index is to ensure that energy supplies are secure. Diversification, whether of fuel type, the geographic location of energy sources, or the companies from which they come, is one of the best ways to achieve energy security. A nation that relies on diverse energy supplies is less likely to be severely impacted by one accident or act of malfeasance (Awerbuch and Sauter 2006, Bahgat 2006, Costantini et al. 2007: 223–24, Dell and Rand 2001, Grubb et al. 2006: 4051). Emphasizing less destructive energy sources can also promote security since people will be able to continue to benefit from its use while experiencing less environmental and social disruption. Two indicators of the economic dimension are supposed to work together to monitor the security of the energy supply by measuring diversity: the fraction of the total primary energy supply comprised of renewable sources and a measure of energy import dependence. Let us examine each in turn.

Increasing the share of renewables in a nation's total energy supply is taken to be a movement toward sustainability because they can improve energy security by 1) increasing the number of types of energy sources in use; 2) increasing the diversity of geographic regions from which energy is obtained; 3) decreasing the amount of greenhouse gases that are emitted; and 4) decentralizing energy sources so that one disruption in service is less likely to be catastrophic. Given the dominance of fossil fuels in the energy mix used today, increasing the share of renewables is a worthy goal.

Since renewables are expected to be a growing, yet still minor share of energy supply in the coming decades, this indicator will continue to be

144 *An ethical examination of sustainability indexes*

appropriate for some time, thus it follows the principle of farsightedness into the medium term future (Energy Information Administration 2006). If, however, we look into the far future, as advised by the principle of farsightedness, to a time when renewables are of medium to large significance in the energy supply, this indicator will be less able to measure the diversity of fuel sources as it lumps all renewables into one category. This potential future limitation in the efficacy of the renewables indicator should not hinder its use today because this scenario is unlikely for decades. By that time, analytical tools that more comprehensively measure security, and the data to employ them, will be available and the index will be able to adapt accordingly.

In the present, the goal of security measured by the renewables indicator aligns with multiple broad principles including justice. Secure energy services are of utmost importance if justice is a goal since the poor 1) are significantly, positively, affected by small increases in energy use; 2) are the first and most severely affected by energy shortages, price spikes, and the negative effects of energy use; 3) cannot afford to pay more for energy; and 4) have less political power to lobby for their interests. Increased energy security also diminishes the likelihood of violence over fuel, a benefit to all people. Additionally, it can help stabilize the economy, which in turn can make jobs, products, and transportation possible. Of course, securing energy services does not entail that just access to the benefits of energy services occurs so if justice is important, it also needs to be directly motivated elsewhere in the index.

Promoting renewable energy sources also encourages the careful use of nonrenewable sources. Such an effort is farsighted for it recognizes that while environmental concerns presently overshadow concerns over the physical supply of fossil fuels, fossil fuels are limited. Preserving some of these resources along with the assimilative capacity of the environment for future generations is the caring and responsible thing to do.

The renewable indicator applies to developed countries and is adaptable to developing insofar as it registers those that are actively transitioning to high quality marketed energy sources. The renewable indicator is limited with respect to adaptability insofar as it does not register nonmarketed biofuels that are renewable and make up a large share of the energy use of billions worldwide. But, as previously mentioned, reliable data about nonmarketed traditional biofuels does not exist and therefore this indicator is fairly adaptable within the parameters of feasibility. In sum, the renewables indicator follows the broad principles fairly well though limited data hampers its ability to align with adaptability and justice.

The final indicator of the economic dimension, the measure of energy import dependence, also aims to focus on the goal of energy security. The import dependence index registers decreasing energy imports as a move toward security and thus toward sustainability. The index was constructed as it was because of its association with two of the largest risks from energy use today: 1) the possibility of war, political blackmail, or interrupted supplies of fossil fuels, particularly oil, because of the actions of hostile countries or groups; and 2) climate change from the use of fossil fuels. It is true that decreasing reliance on imports from hostile or unstable regions can increase energy security. Additionally, moving to domestic energy services often implies using more renewables that, as we have seen, can increase security. In these scenarios, decreasing import dependence is a move toward SED.

Yet, the limits of this approach are obvious: imports are not necessarily bad. Not all countries are so unstable that relying on them for energy is more risky than the security benefits of increasing the diversity of geographic sources and types of fuel that are possible when obtaining energy from other countries. For example, hydroelectricity from Canada used in the United States is a more stable resource than oil from the Middle East given political relationships between the countries. Indeed, importing hydroelectric power from Canada may actually increase US energy security by increasing the diversity of the energy supply and reducing the global warming risk by limiting the need for coal fired power plants. Few countries have enough energy resources within their boundaries to meet their energy demands while reducing environmental impacts and ensuring constant supplies. Therefore, relying solely on domestic energy resources can *decrease* security. Yet, as it stands, the index counts all energy imports as equal threats to sustainability.

While the three-dimensional index's measure of import dependence is a flawed measure given the fact that decreasing imports is far from unequivocally a move toward SED, there are no existing measures that are robust enough to substitute for it in the index. Indeed, experts have highly divergent opinions about the appropriate analytical measures of energy supply mixes (Brower 1995: 115, IEA/OECD 2001: 93, Lucas et al. 1995: 5, Stirling 1994). Thus, the question remains as to whether the indicator ought to be left in, flaws and all, to be replaced with a more thorough indicator at another time, or if it ought to be taken out altogether. I suggest that the import dependence indicator be removed from the three-dimensional index because of its inability to clearly point toward or away from SED. The limitations of depending on energy imports should be registered by decision-makers through means other than the index until analytic methods achieve wider acceptance and deeper rigor.

Fortunately, the renewables indicator is a decent measure of fuel type and geographical diversity given current energy supply mixes. After all, only 8 percent of the world's marketed energy came from renewables in 2003 (Energy Information Administration 2006). Increasing the market share of renewables will certainly support fuel-type diversity and thus security. Renewables also promote security by decreasing climate change and, because they are often decentralized, their use can limit the chance for a major energy supply disruption and subsequently increase security. If implementable indicators of geographic, source, or fuel-type diversity are developed they should be incorporated into the index. Until that time, the renewables indicator will be sufficient.

146 An ethical examination of sustainability indexes

Since I recommend eliminating the energy import indicator from the index because it does not adequately align with respect to the situation on technical grounds, it does not make sense to evaluate it according to the other ethical principles. Unclear and potentially misleading information will not help one live by the ethical principles and, indeed, will only hinder efforts to apply them.

As we have seen, the economic dimension of the SED index has larger problems than either of the other dimensions, yet is still an essential part of any index of SED. I advise removing the import dependence indicator and recognizing that the social dimension can monitor economic strength. With these changes, the intensity and renewables indicators, when paired with the other dimensions, do indicate whether a country's energy use is moving toward increasing energy efficiency and security, two aspects of economic SED. They also align with the principles of careful use, responsibility, far-sightedness, adaptability, and justice about as much as possible given the availability of data and analytical methods. Thus, the three-dimensional index can say something about the sustainability of the economy with respect to energy use, an essential characteristic of sustainability as a whole.

5.2.5 An assessment of the entire three-dimensional index

Now that we have assessed the dimensions of the SED index and its individual subindexes, we can take some time to evaluate its merits as a whole. The most consistent drawback to the index is that potential improvements suggested by the ethical analysis regarding the social and economic dimensions and the principle of justice as applied to all dimensions, are hindered by 1) data availability, 2) methodological constraints, and 3) the desire for a simple index. In the near future, these problems cannot be adequately addressed. In the more distant future, better data will become available and may be able to be incorporated into the index, especially if index developers identify ideals for data and interact with data collection and policy-making.

Although many of the suggestions most recommended by the ethical principles are difficult to implement, at the present time there is merit to noting these suggestions whether or not they are included in the index. First of all, they help delineate the limits of the index so it is recognized for the potentially helpful but incomplete index that it is. Second, many improvements were suggested for both technical and ethical reasons, a fact that should lend support to the call for further research and data collection. Third, when the principle of justice helps us realize that the negative, disproportionate impacts of energy use are not included in the index, we understand the ways in which the SED index falls short of the normative claims of sustainability. All of these insights will help decision-makers pair the index with additional sources of quantitative and qualitative information that will help them fully assess the sustainability of a nation's energy use.

And yet, the ethical assessment of the index does not just point out its weaknesses. For example, concerns for how the index treats developed and developing countries differently often disappear when the long-term and multidimensional aspects of the index are considered. We saw this tendency with respect to the energy intensity indicator and the indicators of the social dimension. Thus, the index generally adheres to the principle of farsightedness as it acknowledges that activities such as increasing electricity use may simultaneously enable movements away from SED in one dimension and toward SED in another. The short-term prioritization of one dimension over another should not obscure the movements either away from or toward SED. As it is applicable to the long-term needs of all nations, the index also generally aligns with the principle of adaptability.

As far as the index promotes knowledge about the sustainability of a nation's energy use it enables responsibility. After all, a root of responsibility is the ability to understand and recognize what we are to be responsible for. Therefore, increasing our knowledge about the state of the situation can put us in a position to act appropriately. Of course, such knowledge does not guarantee that people will take responsibility for their actions; it is merely a precondition for acting responsibly in many cases. It is possible that people can sometimes act responsibly without understanding all of the outcomes of their actions, but waiting for spontaneous environmental activism will not ensure sustainability. Tools such as the three-dimensional index are needed to help identify the ways in which humans are and can be responsible for environmental states. With this understanding of responsibility in mind, it is clear that insofar as the index outlines the most egregious implications of energy use it can help people design policies to avoid them.

The principle of careful use is also fairly well followed by the index in that it prioritizes limiting the harmful effects of energy use. Admittedly, to the extent that the index focuses on decreasing the negative effects of energy use without necessarily decreasing the amount of energy used, the index does not necessarily follow the frugality specification of careful use. It could. This "could" is significant because those who focus on frugality could endorse the index even if they would prefer that it more thoroughly prioritized changing lifestyles to decreasing consumption.

The index follows the broad principle of justice applied to humanity insofar as it monitors whether high quality energy and the social benefits that often go with it are accessible and affordable to people. Yet because it is not able to assess whether the poor have affordable access to energy, or whether the burdens of energy use are proportionately distributed among the population, it still leaves many aspects of justice unaddressed.

In summary, the three-dimensional index of SED is partially, but not completely, aligned with the broad principles. A contributing factor to this simultaneous alignment and misalignment with the broad principles is that the principles need to be balanced against each other, especially if the index is to be implemented. Thus, at this point, the ethical critiques of the three-dimensional index suggest avenues for future development and praise aspects of the current incarnation of the index. Generally it affirms that the index developers followed the broad principles, even if unintentionally, and balanced them within the confines of present knowledge and action. The ethical analysis affirms that the three-dimensional index, especially with a revised imports index and more attention to adaptability and justice in general, is a technically and ethically sound way to assess the sustainability of a nation's energy use given the limits of existing data.

5.3 Prescott-Allen's *The Wellbeing of Nations*

So far the evaluation of sustainability indexes has focused on energy indexes since a narrow subject enables a thorough investigation of the index components in a reasonable amount of space. Yet, energy indexes may not be as advanced with respect to national conditions of sustainability as indexes which aim to be comprehensive: the larger and more diverse indicator sets in comprehensive national indexes may yield more opportunities to meet the ethical principles. To test this possibility, this section looks to Prescott-Allen's extensive sustainability indexes which, while not widely adopted, have had some influence on sustainability monitoring initiatives (Fraser et al. 2006: 117–18). The next section examines the 2012 Environmental Performance Index, a newer index which focuses on policy goals rather than an individual's ideals for sustainability. To streamline these sections, the analyses will focus on the novel aspects of the indexes as compared to the energy indexes.

Prescott-Allen's *Wellbeing of Nations* monitors ecosystem and human well-being as well as interactions between them. At a basic level, his index is comprised of the Ecosystem Wellbeing Index (EWI) and the Human Wellbeing Index (HWI), each of which are aggregates of dozens of indicators. Prescott-Allen also tracks the overall well-being of a nation through plots of the country's human and ecosystem well-being on orthogonal axes. Additionally, he calculates the ratio of the HWI to ecosystem stress (the inverse of well-being) to monitor how much human well-being arises from a nation's stress to the environment (Prescott-Allen 2001: 107–9). Throughout these measures, he aims to equally weight human and ecosystem well-being because human society will not be sustainable for long without a strong environment and because he doesn't think that a strong ecosystem is valuable unless human well-being is possible (Prescott-Allen 2001: 4–5).

The EWI and HWI each have five dimensions. Land, water, air, species and genes, and resource use constitute the dimensions of the EWI while health and population, wealth, knowledge and culture, community, and equity make up the dimensions of the HWI (Prescott-Allen 2001: 7). Each of these ten dimensions is a combination of at least two elements, each of which is monitored by a combination of two or more indicators for a total of eighty-seven indicators. For example, the resource use dimension of the Ecosystem Wellbeing Index is comprised of the elements of energy and materials, waste generation and disposal, recycling, and pressures from the agriculture, fisheries, timber, mining, and hunting and gathering resource sectors. The energy and materials element is comprised of the two subelements of energy and materials. Finally, the energy subelement monitors the lower score of a nation's energy consumption per hectare and its energy consumption per person to track the sustainability of a nation's energy use (Prescott-Allen 2001: 7, 96).

In order to aggregate indicators that are measured in monetary or one of many physical units the EWI, HWI, and each of their components are monitored by performance score. Existing data, international standards, estimated rates of sustainable activity, and expert opinion are used to correlate existing data from each indicator to a performance score of 1 to 100, which is divided into five bands: bad (1–20), poor (21–40), medium (41–60), fair (61–80), and good (81–100). The use of performance scores inserts a significant amount of uncertainty into Prescott-Allen's indexes since twenty-five of his eighty-seven indicators had not yet been the subject of thorough empirical study or policy analysis, so he relied on his personal judgment to set their performance scales (Prescott-Allen 2001: 300).

While the five dimensions of each index are combined using equal weights, the elements, subelements, and indicators are combined using a variety of procedures. Some items deemed to be of equal standing in the hierarchy are averaged. At other times only the highest or lowest value of a series of indicators or elements is incorporated into the index because all are understood to be necessary for well-being such that the worst value is what determines the level of well-being. This variety of value judgments makes his indexes difficult to reproduce (Prescott-Allen 2001: 8–9, 269–75, 300, 307–12).

Prescott-Allen's two major indexes, the Ecosystem Wellbeing Index and the Human Wellbeing Index, can exemplify significantly different ethical principles. For instance, the EWI's indicators all promote careful use in that they register a more sustainable action when fewer resources are used or less pollution is generated. The HWI, however, does not directly prioritize or discourage careful use since, aside from monitoring whether a nation meets its people's basic needs with respect to food and water it does not study resource consumption. Indirectly, its health indicators do relate to environmental disruption as they will be lowered when environmental degradation impacts health but these causes are not emphasized in the indicator about life expectancy and fertility rates. Thus, it is safe to say that the HWI itself is less directly related to careful use, though the EWI certainly exemplifies it well.

With respect to adequate assessment of the situation, both the EWI and HWI do identify many significant elements of environmental and human systems, though the indexes are limited by available data as not all indicators are available for all countries. The EWI and HWI do identify components including culture, materials, and the state of the oceans that Prescott-Allen would like to but cannot directly monitor. Since he leaves room for them in the indexes, if they are adopted they may promote more data collection.

On first glance, intrinsic value supporters may be disappointed with Prescott-Allen's work since he frames his indexes, even the EWI, in terms of human needs (Prescott-Allen 2001: 4, 59–60, 66). Upon closer inspection of the indexes, however, intrinsic value theorists will uncover rich resources for monitoring (and thus valuing and preserving) the state of various environmental conditions when compared to the other indicators studied in this chapter. Specifically, various subindexes of the EWI monitor multiple classes of land, flora, and fauna such that users will be able to know if one place or population is at greater risk. For example, the land protection subindex combines an indicator monitoring the size of protected land in a nation with an indicator that tracks the diversity of land protected (Prescott-Allen 2001: 66–73). An subindex regarding species is made of indicators of the diversity of wild and domesticated animals and plants (Prescott-Allen 2001: 86–93). This specificity enables users to track which kinds of entities are most at risk and thus most in need of care, whether people are valuing them for their own sake or as resources for humans. Certainly, a strident biocentrist may wish to change the weight of some indicators to prioritize otherkind and may not appreciate the fact that humans have an index of their own, but the fact remains that the detailed nature of the EWI's indicators enables a more detailed assessment of the relative vitality of various ecosystem components, and thus a possibility of a richer sense of responsibility and justice. The fact that particular indicators may be supported for a range of reasons reinforces the idea that exact ethical alignment is not necessary to evaluate indicators; the broad principles may be quite useful.

Prescott-Allen's indexes are, however, sometimes limited with respect to adequate assessment since it can be difficult to ascertain exactly what his indicators are supposed to or do reveal because he does not always explain them in enough detail. For example, Prescott-Allen intends the energy indicator to ensure that the rate of energy consumption does not exceed the carrying capacity of the environment. He recognizes that we do not know what the carrying capacity of the ecosystem is and that we do not know the extent of environmental damage caused by energy use (Prescott-Allen 2001: 95–96). Given this uncertainty, he monitors two measures of energy use in order to monitor energy's effects on natural and environmental scales. He claims that:

> Energy consumption per hectare of total area is valid if the effects of a country's energy use are absorbed largely by the ecosystems within its borders. Energy consumption per person is more defensible if national energy use has a global impact. Since either may be the case, both indicators are employed here.
>
> (Prescott-Allen 2001: 96)

Yet Prescott-Allen does not explain why energy consumption per person is more applicable for monitoring global impacts of energy use. Thus, it is difficult to understand whether it does follow the principle of adequate assessment.

Admittedly, these impacts could be explained as follows. When a country's energy use largely influences its own land, as when it relies on hydroelectric dams which flood rivers located solely within its borders, tracking the country's energy consumption per hectare over time will provide a measure of the changing impact of energy use on its land. As there is a higher demand for energy, more dams will be constructed and they will negatively influence the land. Thus, the energy used per hectare may indicate something about the environmental effect of energy use within a country's borders.

Energy consumption per capita may be considered a better measure of the international impact of energy use if one thinks that a country's contribution to global environmental destruction and conservation should be proportional to that country's share of the world population. If such a distribution schema could be ensured, there would be more equitable access to the goods and services possible with energy use and the externalities that result from energy use on a national level. (Of course, equity on an individual level would not be ensured.) Until such a goal is reached, comparing the energy use per capita of various countries enables the identification of countries that use relatively low levels of energy per capita yet have relatively high levels of well-being. The policies of such nations can serve as models for countries that use relatively high levels of energy per person with little improvement in well-being. If this is what Prescott-Allen intended then monitoring the energy use per hectare and per capita enables the EWI to monitor the national and international impact of a nation's energy use.

The explanation given above seems the most reasonable interpolation of Prescott-Allen's assumptions. Yet his energy indexes may not meet his goals. His energy indicators can certainly help identify countries, namely those who have low energy use but a high EWI and HWI, whose rates of energy use are an example for the rest of the world. They also help ensure that one can compare energy use between countries of various sizes and populations. Indicators of energy use per hectare and per capita, however, are not necessarily well aligned with national and international impacts of energy use. Countries may use relatively small amounts of energy per hectare yet their energy use may have a large international impact due to greenhouse gas emissions, exports of nuclear waste, or the effects of extracting fossil fuels. Similarly, a nation with high energy use per person may influence the environment most significantly within its own borders if it primarily relies on hydroelectric power generated within its own country. Consequently, while these indicators can be useful, it should be noted that their utility comes from the assumption that lower energy use in general is advantageous, not because the indicators necessarily track national or international effects of energy use. Thus, it is unclear that Prescott-Allen's energy indexes are actually an adequate assessment of the impact of energy use beyond rough measures of consumption.

Additionally, anyone using Prescott-Allen's indicators of energy use per capita or per hectare must recognize that they rely on a number of significant assumptions. First, they assume that population, land use, and energy use data is reliable and readily available. While these three data points are some of the most often collected, this data is still far from accurate, especially in developing countries where nonmarketed energy sources are common but often left out of national energy accounts. Second, Prescott-Allen's energy indicators classify all types of energy use as equally destructive to the environment and equally beneficial to humans. Yet we know that this is not the case. Old coal-fired power plants with and without scrubbers have very different environmental impacts. Similarly, access to high quality energy may be affordable and reliable in a wealthy neighborhood whereas it may be prohibitively expensive and unreliable elsewhere. As discussed in the assessment of the three-dimensional index, data to monitor such nuances of energy use are not yet available. Thus, Prescott-Allen's energy indicators are limited in that they rely on insufficient data and because they cannot register important variations in the effects of energy use.

Admittedly, Prescott-Allen tries to make some of his indicators applicable to countries along the developing–developed spectrum. For example, in his index of water pollution he tracks the major water pollutants in both developing and developed countries in both rural and urban areas: microbial pollution from animal and human waste, acidification, heavy metals, arsenic, nutrient levels from farmland and municipal waste, and oxygen balance (Prescott-Allen 2001: 75–77). Despite this desire for inclusiveness, his indexes are skewed toward the conditions in cities and developing countries because more, and more reliable, data is available in these areas. We especially see this trend in the air pollution indicators which are based on city pollution (Prescott-Allen 2001: 85).

In sum, limited explanations of Prescott-Allen's indicators as well as his reliance on aggregated and limited data reduce the ability of Prescott-Allen's index to align with adequate assessment of the situation.

The indexes fair better with respect to farsightedness. The EWI is farsighted in that it examines environmental effects that have long temporal and spatial scales such as climate change and records decreases in pollution or environmental degradation as a good regardless of a country's current state of development. Additionally, since the indicators regarding resource use aim for national self-sufficiency they indirectly work to ensure that other nations do not lose resources or gain pollution for the country under study (Prescott-Allen 2001: 97–107). Similarly, the HWI is farsighted in that it values birth rates that are at or under the replacement rate, intending to limit pressure on the environment for future generations (Prescott-Allen 2001: 23–24). Beyond this, it tends to focus on the current generation, assuming that current conditions of wealth, knowledge, good governance, peace, and equity will also improve future conditions.

Prescott-Allen's indexes promote responsibility in that they aim to extend knowledge about the state of human and ecosystem well-being, encouraging people to recognize when they have been responsible for degradation and when they could effect change. Additionally, as his work enables comparisons between nations it enables the identification of countries that can serve as examples for others.

Prescott-Allen addresses justice in his indexes in some unique ways though he still has a limited perspective on equity. He only considers justice within the HWI. Here equity appears as a separate subindicator, the average of an indicator of household equity (the ratio of the richest 20 percent of the country's income to that of the poorest 20 percent) and gender equity (the average of the performance scores for the ratio of male income to female income, the gender distribution of education, and gender distribution in the national parliament). Notably Prescott-Allen also maintains that "Income share is the best available indicator of household equity but does not necessarily reflect other areas of potential disparity such as health, knowledge, power or freedom," a significant move to encourage the collection of more data and development of new methods, though these factors are not actually in the index (Prescott-Allen 2001: 51).

Prescott-Allen also includes a measure of whether a country meets its people's basic needs, monitored by the lower of scores for food sufficiency (itself monitored by the lower of the percentage of children with stunted growth, low birth rates, and underweight), and basic water and sanitation services (Prescott-Allen 2001: 27–29). In theory, this approach aligns with justice because it focuses on the worst off, not just on average conditions. In practice, however, it is severely hindered by data limitations, especially in developing countries which do not have infrastructure to provide basic services. Furthermore, Prescott-Allen does not encourage directly monitoring environmental injustice or the unjust distribution of environmental burdens within a population, for example along gender, ethnic, or racial lines. Thus, even though he includes an explicit measure of equity and encourages some further research in this area he is hampered by data limitations and his perceptions of environmental justice.

As mentioned above, Prescott-Allen's relatively detailed indicators enable comparative studies of the disparities in the experience of different types of plants, animals, and ecosystems. Thus, though he himself does not prioritize justice between different biota other than humans, his indicators could be used to investigate such phenomena, especially if disaggregated from the index as a whole. Thus, biocentrists, intrinsic value supporters, and anyone interested in the specific ways biota and ecosystems are or are not thriving may find his indicators helpful, even if for reasons he didn't quite intend. Of course, as always, such investigations are limited by the specificity of the indicators and the reliability of the data; most of these indicators aggregate data quite widely (all domesticated animals, all wild mammals etc.).

Prescott-Allen's indexes align with adaptability in a variety of ways, but here too, are limited by data availability. Subindexes in both the HWI and EWI are often composed of many indicators to track the significant aspects of many different nations' adherence with that particular aspect of well-being. For example, three main types of plants (flowering plants, gymnosperms, and ferns) are tracked, to the extent that data are available, in order that the wild plant diversity subindex is thorough and applies to countries with very different ecosystems. (Prescott-Allen 2001: 87–91). Innovatively, Prescott-Allen's indexes sometimes use different specifications of an index for different nations. Specifically, the indicator of domesticated animal diversity tracks the number of breeds for each of the top three most popular types of domesticated livestock in a nation. Thus, one nation's index may track breeds of cattle, sheep, and goats while another may track camels, horses, and goats. In this way, the indicators allow the particular situations of each nation to be monitored while tracking the overall diversity of domesticated animals. Unfortunately, statistics regarding domesticated animal diversity are quite poor and rarely compiled, so the applicability of this indicator is limited (Prescott-Allen 2001: 91).

With respect to feasible idealism, Prescott-Allen tends toward idealism because the complexities of his index and the temporal and monetary cost of applying it prevent it from feasible implementation. Indeed, one project based on his work took an extra year to generate results because of the timeframe of getting the data, the complexity of the results, and the distance of the results from the decision-making process. In the end, the index was not even used because it was too complex and took too long (Fraser et al. 2006: 117–18). Prescott-Allen's indexes blend feasibility and idealism in that he articulates goals for the indexes whether or not data is available and then measures whatever he can get his hands on, developing methods for dealing with the fact that different sets of indicators are available for different countries. For example, Prescott-Allen calculates subindex values even if individual indicators are missing, though he limits a nation's subindex performance score if it is missing indicator data because he does not want countries to appear to be doing well if data is not available (Prescott-Allen 2001: 9). This method enables him to calculate index results for more countries, giving them a baseline even if all of the data is not yet available. If his methods are adopted, this process may also push data collection in a variety of areas. However, the complexity of Prescott-Allen's indexes and the fact that they are based on so much probably unreliable or unavailable data makes them less feasible to implement and yield unreliable, uncertain results if they are implemented.

In sum, Prescott-Allen's index aims for comprehensiveness and has features that begin to align with the broad principles, but its alignment with nearly every ethical principle is limited by data availability, the complexities of the index itself, and its disassociation from the policy-making process. His two main innovations compared to the energy indexes relate to adaptability and justice. First, buried in the index is the fact that he allows nations to monitor the three forms of livestock most significant in their country rather than those on a predetermined list, a way to recognize the diverse agricultural conditions of nations. Second, he recognizes the importance of social equity for well-being, a significant move toward a more thorough method of monitoring justice, though directly connecting it to environmental issues would be even more relevant.

5.4 The 2012 Environmental Performance Index (EPI)

As we have seen, the largest challenges when constructing sustainability indexes are the lack of data with which to measure the index, uncertainty about sustainability targets, and the distance of the indexes from justice. The Environmental Performance Index (EPI) addresses these issues by measuring progress toward widely agreed-upon policy targets such as the United Nation's Millennium Development Goals (MDG) which aim to reduce poverty, support growth, and facilitate sustainable development by 2015. By doing so it relies upon the most studied indicators and the new data gathered because of the MDG. Indeed, the index has been updated every two years since 2006 to take advantage of new data. Additionally, the 2012 index uses the most robust international time-series data to not only yield a snapshot assessment of a nation's performance but also study trends within nations over time (Emerson et al. 2012a). Ethically speaking, the index emphasizes ecosystem vitality and indicators which correlate with the health of the poorest yet it, like the other indexes, is rarely able to monitor variations in environmental or social conditions within a nation.

The EPI was developed by members of the Yale Center for Environmental Law and Policy and the Center for International Earth Science Information Network at Columbia University in collaboration with the World Economic Forum and the Joint Research Centre of the European Commission. They have intentionally developed the EPI to track "the ability of countries to actively manage and protect their environmental systems and shield their citizens from harmful environmental pollution" while encouraging environmental performance which "promotes action, accountability and broad participation" rather than monitoring sustainability comprehensively. Indeed, they focus upon the specific issues for which national governments can be accountable based on existing policy (Yale Center for Environmental Law and Policy et al. 2010: 64–65). By doing so, they aim to avoid the difficulties that arise from variations in sustainability definitions and more effectively guide policy-making (Emerson et al. 2012a: 11). While the EPI developers distinguish their index from a full sustainability index, I include it in this analysis because it assesses preconditions for sustainability.

Two subindexes, Environmental Health and Ecosystem Vitality, comprise the EPI. Environmental Health includes indicators of environmental degradation of human health (represented by childhood mortality rates), indoor and outdoor air pollution that directly affect humans, and access to water and sanitation services. Ecosystem Vitality tracks the status of the ecosystem itself and the ways humans manage natural resources through indicators of climate change, agriculture, fisheries, forests, biodiversity and habitat, water resources and the effects of air quality on ecosystems. Many of these subsections are comprised of a combination of indicators, for a total of twenty-two (Emerson et al. 2012a: 16, 21–23). Like the indexes evaluated elsewhere in this chapter, the EPI tends to emphasize human activities, experiences, and perspectives, though its indicators of ecosystem vitality could be the basis of valuation and support of ecosystems in and of themselves.

With this general overview of the 2012 EPI we can now turn to the ethical analysis of the index, emphasizing aspects that are ethically distinct from previously analyzed indexes. The regular and trend EPI will not be differentiated in this analysis since they only differ with respect to whether or not they are based on time series data.

Of course, the 2012 EPI is limited with respect to assessment of the situation because of its intentional focus on human health impacts from the environment and ecosystem vitality rather than the whole scope of sustainability. The index developers advocate focusing on these preliminary conditions for sustainability as they seem quite important and are agreed upon in policies. Additionally, it is important to note that they are not exclusively focused on the most direct human and ecosystem health as when they look for correlations between the indicators of the 2012 EPI and GDP, finding that GDP is correlated with a decreased environmental burden of disease (Emerson et al. 2012a: 9).

While advantageous in many ways, the 2012 EPI's focus on policy goals can be limiting as measures of human health or ecosystem vitality are not necessarily complete enough to ensure sustainability. There may well be important environmental initiatives not yet codified into policies and so will not be included by the 2012 EPI. In the short run, such a focus may well help achieve environmental goals. In the long run, however, those who use the 2012 EPI must not forget that it is limited to existing policy developments. Because the 2012 EPI developers are explicitly open to revising the index as new data and analytical tools are available and as policy targets are articulated, and indeed, do so every two years, it is reasonable to think that they would also be open to revising the index as new policies are developed.

Continuing the evaluation of the 2012 EPI according to adequate assessment of the situation we see that the developers of the 2012 EPI repeatedly acknowledge that the limited availability and quality of data hinders its thoroughness, especially since it only includes indicators for which robust, time series data is available. For instance, several dozen countries are not assessed because data is not available for the index components. The index developers would also like to monitor

> "toxic chemical exposures; heavy metals (lead, cadmium, mercury); municipal and toxic waste management; nuclear safety; pesticide safety; wetland loss; species loss; freshwater ecosystems health; water quality (sedimentation, organic and industrial pollutants); recycling; agricultural soil quality and erosion; desertification; comprehensive greenhouse gas emissions; and climate adaptation"

for which there is not yet sufficient robust, international, time series data (Emerson et al. 2012a: 15). Even where data is available it has limits, as when satellite data is used to measure particulates in the air. There may be significant but as yet undeterminable differences in air quality between ground level and higher air and even this data is only collected to 60 degrees North, excluding parts of Norway, Russia, and other high latitude countries (Emerson et al. 2012a: 37). These measurements are, however, a considerable improvement over air quality measurements used in previous indexes which relied upon measurements in a few cities per country (Yale Center for Environmental Law and Policy et al. 2010: 38–39).

Earlier versions of the EPI, especially the 2010 version, were able to capitalize on data available for the first time to overcome prior data gaps. For example, more data about nonmethane volatile organic compounds and NOx, important air pollutants, as well as signatories to international regulations about pesticide use, an indicator of pesticide use, were included (Yale Center for Environmental Law and Policy et al. 2010: 35, 41). The 2012 version, with its increased data standards, has dropped some of these indicators. The index developers are, however, open to the inclusion of new indicators in future indexes as better data become available. Indeed, they identify their next task as the improvement and implementation of measurement schemes on one issue at a time, starting with "air quality and pollution" (Emerson et al. 2012b).

Thus, the 2012 EPI narrative is very forthcoming about its limitations with respect to data and methodology, where it has improved, and where it aims to improve in the future. These trends not only exemplify the principle of adaptability but also illustrate that the index carefully assesses the situation with respect to the best available data and that its developers aim to make it transparent to aid policy-makers and thus increase the feasibility of its use.

The index is also adaptable insofar as its basic calculation methods and particular indicators enable its application to a wide range of countries. For instance, the 2012 EPI uses a log scale for many indicators to distinguish nations near the target of an indicator and between these nations and those far from the target (Emerson et al. 2012a: 17). Three indicators of greenhouse gas emissions monitor emissions common in various nations. For instance, measures of greenhouse gas emissions per capita include emissions from agriculture and changes in land use that are often most significant in developing countries. The second indicator monitors CO_2 emissions per unit of electricity generation, the major contribution of anthropogenic greenhouse gas emissions for many industrialized nations. Finally, the EPI monitors CO_2 emissions divided by kilowatt hours of electricity used to encourage less carbon intensive methods of energy generation (Emerson et al. 2012a, Yale Center for Environmental Law and Policy et al. 2010: 60–61).

The 2012 EPI aligns well with the principle of justice insofar as it focuses on the environmental contributions to health. While in previous years it monitored the environmental burden of disease – the decrease in healthy years and decrease in life expectancy because of environmental conditions – the 2012 EPI uses childhood mortality (up to age five) as a measure of the health impacts of degraded environments. This measure is more readily available and robust than measures of the environmental burden of disease, yet captures whether one of the most vulnerable populations, young children, are killed by their environment. Additionally, it focuses on water quality and air pollution, the major contributors to the most significant environmentally related diseases. By concentrating on environmental indicators most connected to disease, the index examines the direct results of degraded environments on human health rather than assuming that degraded environments are problematic, as other indexes do. Since these diseases are mainly experienced by the poor, the index also steps toward monitoring distributive justice by focusing on these problems.

Limitations to the indexes' work on justice include the fact that childhood mortality may be significantly influenced by good health care systems, masking the country's environmental disease burden. Additionally, because the index focuses on environmental health and environmental conditions it does not monitor participatory justice or the ability of people to sustain important aspects of their culture. Finally, as in the other indexes, the 2012 EPI generally does not enable index users to examine disparate environmental burdens within a nation. Thus, while the 2012 EPI does target social dimensions of sustainability and justice insofar as it focuses on the direct effects of environmental degradation or human health, especially those experienced by the youngest and poorest people, it leaves room for improvement with respect to the principle of justice.

The 2012 EPI balances idealism and feasibility quite well. It prioritizes relevance to policy-making so that it is more likely to be used. It expresses its ideals as it notes how it can be improved upon when more data is available and specifies what types of data should be collected. While it can be idealistic because there is often a long way to go to reach its targets, since there is usually widespread political agreement about its goals, it is more realistic that there will be slow movement toward them. For indicators which do not have agreed-upon goals, for example for reducing greenhouse gas emissions per unit of electricity generated, the index developers set ambitious goals, in this case zero, because high goals enable comparison while pushing everyone to do better (Emerson et al. 2012a).

Setting such long-term goals also makes the index quite farsighted in that it will not be obsolete for some time. Additionally, as the greenhouse gas emissions subindex comprises 25 percent of the EPI, it focuses on the long-term and far-reaching consequences of human activity. The index is also farsighted as it sets the target for greenhouse gas emissions to 50 percent of 1990 levels by 2050, a likely target for policy, but one that is still a ways from being accepted potentially; and even farther from being realized (Yale Center for Environmental Law and Policy et al. 2010: 60).

All of these trends, particularly the focus on pre-existing policy targets, enable users of the 2012 EPI to take responsibility for their actions by assessing outcomes of previous policy initiatives, identifying countries with best practices within peer groups, and enabling the development of new initiatives. In sum, the 2012 EPI is most closely aligned with the ethical principles of the indexes examined so far. It makes strides with respect to adaptability, farsightedness, and justice, though there is still room for the EPI developers to improve ethically in later iterations because of its limited ability to discriminate between the experience of subpopulations within nations.

5.5 Eurostat's Sustainable Development Indicators (SDI)

Examining the degree to which the carbon emissions index, three-dimensional index of sustainable energy development, the Wellbeing of Nations, and the 2012 EPI align with the ethical principles yields a general sense of how well sustainability ethics are incorporated into indexes. None of these indexes, however, is an official mode of monitoring progress toward or away from sustainability and thus their influence on policy-making and action is less than certain. On the other hand, we have the Sustainable Development Indicators (SDI) developed by Eurostat, the office which aggregates statistics for the European Union, which directly monitor and influence sustainability initiatives. Examining this official list of indicators for consonance with the ethical principles is thus critical for understanding the degree to which the ethical principles are involved in official sustainability initiatives, and how sustainability can be monitored with available data. This analysis will reveal that the Eurostat SDI include several innovations with respect to adequate assessment of the situation, farsightedness, and justice compared to the indexes examined above. They are, however, still limited by available data, particularly with respect to distributive justice and participatory processes of index development.

According to the EU, democracies require data to inform decision-making by leaders and the public at large. Thus Eurostat is charged with "provid[ing] the European Union with statistics at the European level that enable comparison between countries and regions." Eurostat does not collect data itself; that task is left to member countries of the EU. Rather "Eurostat's role is to consolidate the data and ensure they are comparable." (Eurostat 2012a). While Eurostat collects basic data on economic, social, industrial, environmental, and technological issues among others, (Eurostat 2012b) it also collects statistics regarding policy themes of the EU including at least 130 "Sustainable Development Indicators" (SD) which align with the EU's Sustainable Development Strategy (SDS) (Eurostat 2012b). Having free access to this information through the Eurostat website fosters participatory justice and allows laypeople and decision-makers alike to better take responsibility for their actions as they come to understand their effects.

Eurostat identifies eleven indicators as headline indicators which together yield an overview of a country's sustainability. This approach mediates between the possibly overwhelming list of 130 SDI and the problems of aggregation if a single index is developed. Thus, Eurostat focuses on the feasibility of implementing its SDI while still enabling the development and use of an extensive list of indicators which approximate an ideally comprehensive list. While a broad picture of a nation's movement toward (or away from) sustainable development using a few indicators is valuable, here I examine the whole set to determine the degree to which it is possible to monitor progress toward the ethical principles using the SDI.

Since Eurostat is primarily a statistical office that happens to monitor sustainability it is quite concerned with the reliability of data it uses. Thus, Eurostat's website includes not only the data for each indicator but also a "quality profile" for nearly every indicator. These profiles define the indicator, the unit or units responsible for collecting its data, the European legislation relevant to the data, and the "objective and relevance of the indicator." They also present an assessment of the "accuracy and comparability" of the data over time and between countries. Indeed, Eurostat grades its data according to a four-level system and notes whether its comparability between countries and over time is high or "restricted" (Eurostat 2012d). Indicator profiles also identify ways the indicators could be improved and acknowledge areas for which indicators still need to be developed. For instance, Eurostat is actively working to improve natural resource indicators (Eurostat 2012d). Through this detailed analysis of the quality of its data and indicators, Eurostat and its sustainable development indicators explicitly align with adequate assessment of the data.

While this focus on high quality data in Eurostat does yield strong results and can encourage the development of new indicators and the collection of more robust data, it can also limit the scope of sustainable development monitored by Eurostat. For instance, Eurostat has a headline indicator for each of its themes of sustainable development except for "good governance," the "coherence between local, regional, national and global actions in order to enhance their contribution to sustainable development" (Eurostat 2012d). Maybe this is because Eurostat has not yet identified an indicator to represent an overarching view of this multifaceted theme, maybe it is because the data is not yet available. In any case, this facet of sustainability is not represented in the list of headline indicators and thus good governance is effectively deprioritized among the indicators. Eurostat's vision of sustainable development is also limited as there are features of sustainable development which it does not even acknowledge in its narratives, let alone in its indicators. For instance, while Prescott-Allen acknowledges that disaggregating data about human quality of life according to a variety of demographic factors will yield a more detailed picture of whether equity is being achieved, Eurostat does not admit the need to expand upon the demographic features it monitors (Prescott-Allen 2001: 51). Thus, Eurostat's SDI will be limited in their ability to monitor sustainable development in a variety of situations unless the EU, and thus Eurostat, significantly changes its perspective. Additionally, some will find the SDI limited in that they, and the rhetoric used to discuss them, emphasize human perspectives and experiences, though some indicators of environmental quality such as the common bird index (Eurostat 2012d) can be interpreted to discern and promote the conditions of otherkind.

Other indicators of the Eurostat SDI set are also less than ideally adaptable to different conditions within the EU. For example, the indicator of fish catches beyond "safe biological limits" only includes fishing in the North East Atlantic, excluding monitoring biodiversity in other places and disparate access to biota within and between nations. Admittedly, this limitation arises in part from a lack of data, especially reliable consistent data across EU Member States, and Eurostat is working to improve the fish catch indicator and the data it relies upon (Eurostat 2012d). Yet currently, the SDI, including the fish catch indicator, is not consistently applicable to the conditions of a variety of nations in the EU.

Though the Eurostat SDI are not yet equally applicable to all EU conditions, they do embody adaptability in a variety of ways. For instance, Eurostat aims to shift personal transportation away from automobiles and toward other modes including rail and inland water transportation for a host of reasons. Personal vehicle transportation and road transportation for freight are the least energy efficient mode of transportation, produce the most emissions of all sorts of transportation, cause the most accidents, and have major economic impacts. The indicators which monitor changes in transportation types include many modes of transportation to capture the infrastructural capacities of a variety of countries (Eurostat 2012d). The SDI are also adaptable to a wide variety of conditions within nations insofar as they monitor poverty risk after social transfers (Eurostat 2012d), thus monitoring the level of poverty risk experienced by members of nations after their nations' welfare programs kick in.

The Eurostat SDI are also adaptable to the conditions of a wide variety of nations in their alignment with the principle of careful use. They prioritize decreasing the amount of materials used in consumption and production as well as decreasing pollution. A number of indicators focus on decreasing the negative effects of using materials. Some prioritize decreasing fossil fuels since they are major contributors of greenhouse gases and climate change. Others trade increases in the proportion of electricity generated from renewable sources, the proportion of fuel generated by renewable sources, or combined heat and power generation production. Indicators which directly monitor pollution include indicators of municipal waste, hazardous waste, sulfur oxides, nitrogen oxides, nonmethane volatile organic compounds and ammonia emissions. Eurostat also monitors indirect means of achieving careful use through indicators including the percent of land that is organically farmed and participation in voluntary ecolabling or ecomanagement and audit schemes (Eurostat 2012d). While these commitments may directly correlate with decreased use of natural resources they can also indirectly promote careful use in the future since they provide more

162 *An ethical examination of sustainability indexes*

information to consumers and can increase the demand for sustainable goods and services.

Indeed, the SDI set is farsighted in a number of ways, some of which duplicate other indexes' alignment with farsightedness and some of which are new. Like the other indexes examined, the SDI tracks environmental and social conditions that, if unchecked, will significantly impact those far away in space and time. The Eurostat SDI add indicators of energy intensity, fossil fuel dependence, renewable energy, and bird and fish populations (Eurostat 2012d). Without such resources, the poor and marginalized alive now and future generations in general will have fewer environmental opportunities compared to the well-off in the present. Thus these indicators can foster farsightedness. Additionally, when it discusses poverty indicators, Eurostat recognizes that poverty can have a long-term effect on families as future generations' health, job prospects, education, and ability to be integrated into society is influenced by their ancestors' economic status (Eurostat 2012d). While many other indexes have monitored poverty to some degree, Eurostat explicitly frames poverty as a present and future-oriented issue. Similarly, the SDI monitor the economic conditions and employment of those over sixty-five years old (Eurostat 2012d), recognizing that if the workforce is overly burdened by caring for the elderly it will not be economically sustainable.

The Eurostat SDI also align with spatial farsightedness to some degree as they try to track the effects of the nation being studied on other nations, data that can help ensure that one nation's progress toward sustainability does not come at the expense of another. For example, it monitors "the amount of official development assistance given to other countries as a share of GNP, recognizing that aid may be necessary for improving and ensuring basic quality of life and global solidarity" (Eurostat 2012d). While there are other ways that one country can influence another, monitoring development assistance is a way of recognizing the relationships between nations. So far, the SDI include fewer indicators of the international environmental impact of the monitored nation. The SDI do aim to ensure that economic growth within a country can continue without consuming ever increasing amounts of natural resources by monitoring the degree to which the economy is decoupled from the use of raw materials and environmental destruction (Eurostat 2012d). They do not, however, count the natural resources indirectly utilized by the studied nation, hindering spatial farsightedness. A nation could appear to move toward sustainability while causing other places to move away from it if it shipped its waste elsewhere or relied on polluting manufacturing elsewhere. While Eurostat acknowledges this limitation of its indicators, it is also unable to rectify the situation given current methodologies, data, and the fact that the SDI focus on individual nations.

Similarly, the SDI's relationship with justice is one of partial alignment. The connection between the SDI and justice is firmly rooted in the EU Sustainable Development Strategy (SDS) to "improve ... the quality of life for present and future generations" (Eurostat 2012c). At minimum, the EU desires distributive justice through increasing "quality of life"

> Sustainable development ... stands for meeting the needs of present generations without jeopardizing the ability of futures generations to meet their own needs – in other words, a better quality of life for everyone, now and for generations to come. It offers a vision of progress that integrates immediate and longer-term objectives, local and global action, and regards social, economic and environmental issues as inseparable and interdependent components of human progress.
> (European Commission 2012)

This emphasis, when linked to the EU's commitment to democracy, suggests that both distributive and participatory justice should be well represented in the SDI.

Eurostat experts acknowledge that present and future generations are important and deserve distributive justice when they track current conditions such as the life expectancy of newborns, poverty, access to work, and education not only for what they indicate about the present but also for what they indicate about the future (Eurostat 2012d). Similarly, they step toward monitoring equitable distribution between current and future generations as they prioritize decreasing greenhouse gas emissions, energy intensity, and fossil fuel dependence as well as increasing indicators of renewable energy, the economic conditions of those over sixty-five years old, and biodiversity (Eurostat 2012d). While they recognize that improvements in these indicators do not guarantee intergenerational justice, they can be steps toward it (Eurostat 2012d).

In contrast with some of the other indexes and indicator sets, the SDI also, to some extent, monitor distributive justice between demographic groups. Usually they compare conditions for males and females as when they track differences in life expectancy, employment, and poverty risk according to gender. Sometimes demographic disparities are tracked for other groups including "age (unemployment, risk of poverty), level of education (employment rate, risk of poverty), household type (risk of poverty) and geographic region (dispersion of regional GDP per inhabitant)" (Eurostat 2012d). Such detail enables people to know when particular groups experience higher risks, knowledge that can prompt investigations of the cause of such disparities and initiatives to limit both disparities in risk and overall risk.

Yet many SDI cannot recognize disparities within a nation because they rely upon average data (Eurostat 2012d). For example, indicators of environmental degradation are not disaggregated by demographic factors or geographic regions within a nation. Thus, the SDI cannot track whether some people in a nation are more likely to be harmed by environmental damage than others. Similarly, aside from the indicators mentioned above, the SDI usually do not monitor differences in social and economic conditions among demographic groups (Eurostat 2012d). Of course, limited data, particularly the need for consistent and reliable data, contributes to these conditions. Yet Eurostat rarely even acknowledges that further disaggregating its data could be beneficial, so such moves are unlikely in the near future (Eurostat 2012d).

While links between justice and sustainability often focus on the distribution of material resources and environmental harms, participatory justice may also be important. Indeed, the SDI set aligns with participatory justice in three broad ways. First, it makes data freely available, enhancing the possibility of informed participation in decision-making. Second, indicators of the "proportion of the population living in households considering that they suffer from noise" and the level of citizen's confidence in EU institutions are based upon survey data (Eurostat 2012d). Their existence indicates that at least some Eurostat experts acknowledge that public opinions on matters such as social and environmental conditions are critical to sustainability as well as standardized "objective" measures. For instance, environmental degradation or a government perceived as untrustworthy may not only have direct physical or economic effects but also lead to stress which in turn may cause physical and mental health problems or a degraded community. The Eurostat SDI set also includes indicators which monitor public participation in government and the openness of the government (Eurostat 2012d). Such indicators include measurements of voter turnout rates, the availability of online government services, and the rate of online services as well as citizen's reported confidence in EU institutions. Finally, because Eurostat's indicators focus on policy targets determined through democratic political processes, the public is indirectly involved in indicator formation. It could, however, be significantly strengthened to ensure that priorities of the public are well represented in the indicators.

As we have seen, the SDI move toward distributive and participatory justice in a number of ways. They generally support responsibility, careful use, and feasibility. The SDI also align with adequate assessment, and farsightedness. Indeed, their attention to the reliability of their data and indicators, their attempts to monitor the economic effects of the nation being studied on other nations and indicators of community perceptions of environmental quality (noise) and trust in government represent advances in comparison to other indexes studied. Once again, data limitations and averaged data significantly limit the indicators. Furthermore, since the development of the SDI (and all of the indexes examined so far) relies heavily on expert indicator developers, it does not facilitate participatory justice as much as if the public was directly involved in the process. Local indexes and indicators, as discussed below, often aim to overcome this challenge.

5.6 Local indexes

Many local sustainability indexes have been influenced by the initiative to implement Agenda 21 on a local level (Fraser et al. 2006: 122, McMahon

An ethical examination of sustainability indexes   165

2002: 185). Such local indexes, like national indexes, can examine sustainability as a whole or an aspect of life (they are often partial to agriculture). In theory, local indexes can overcome some of the participatory justice critiques of national indexes because they are often developed with local input about values and social and environmental conditions. As we shall see, local indexes often do have a richer social dimension than national indexes and embody participatory justice in their method of development but are often hindered by data limits, their narrow concepts of locality, and the fact that methods of monitoring justice are still under development.

Of the hundreds of local indexes, I chose a representative sample among those which emphasize social sustainability to draw conclusions about this genre of index without replicating the same analysis as in the national indexes. Selected indexes monitor many aspects of sustainability including agriculture, forestry, energy, and general sustainability and were developed for communities of a variety of sizes in the UK, the EU, Canada, Mexico, Latin America, Thailand, China, and the Kalahari (Fraser et al. 2006, Gallego Carrera and Mack 2010, Lopez-Ridaura et al. 2002, McMahon 2002, Praneetvatakul et al. 2001, Reed et al. 2008, Wei et al. 2007). Given my focus on social sustainability, the local indexes and indicator sets emphasized the human relevance of the environmental systems that they monitored. Community members participated directly or indirectly in the index formation. In some cases, they participated from the beginning (Fraser et al. 2006, Lopez-Ridaura et al. 2002); in others, they were included over time (McMahon 2002, Wei et al. 2007); and in one case, experts predicted community priorities (Gallego Carrera and Mack 2010). Within the selected indexes and indicator sets, I emphasize unique indicators compared to the national indexes discussed above.

When considering whether these local indexes and indicator sets fulfill the principle of adequate assessment of the situation, one thing is striking: their sheer novelty of indicators. While they may include common measures such as water pollutants, greenhouse gas emissions, education, and GDP, they often use indicators particular to the place in question. For instance, agricultural sustainability indicators for the Kalahari monitor the presence of plants attractive to livestock (Fraser et al. 2006, Reed et al. 2008). They may also examine small-scale environmental issues including household waste and recycling rates, "complaints of dog fouling," access to public transportation (Fraser et al. 2006: 123, McMahon 2002: 179), and local noise levels (Fraser et al. 2006: 123, Gallego Carrera and Mack 2010: 1033, McMahon 2002: 179). Social issues are often represented in more detail in local indicator sets, some of which monitor "attendance at cultural events or access to green space, subjective experiences of health, fear of crime, or the percent of the population dissatisfied with the neighborhood" (Fraser et al. 2006: 123, Gallego Carrera and Mack 2010: 1033, McMahon 2002: 180) or the connections between "political stability and legitimacy," "quality of life," or social risk to specific energy sources and energy use in general (Gallego Carrera

166 *An ethical examination of sustainability indexes*

and Mack 2010: 1033). Local indicators sometimes also combine community and scholarly expertise as they identify indexes. In this way they extend the parties involved in adequate assessment of the situation (Reed et al. 2008). All of these indicators and methods presume that local indexes can and should reflect local priorities and social and environmental conditions.

As I selected these indexes for their attention to social issues, it should not be surprising that they monitor them in more detail than environmentally focused local indexes or national indexes. Yet, local indexes still rarely monitor the disparities within their community with respect to benefits and burdens of environmental disruption for humans. Additionally, the indexes do tend to have such a narrow vision that they do not monitor the way their community affects or is affected by other communities, regions of the world, or biota. The major exception to this trend occurs in indexes which aim to be comprehensive and monitor greenhouse gas emissions (McMahon 2002: 179). Thus, local indexes significantly advance thinking with respect to adequate assessment of the situation, though their assessments still face significant limitations.

Similarly, the ability of these indexes and indicator sets to follow the principle of adaptability is both helped and hindered by their focus on particular places. This narrow focus enables the indexes to be aligned with local values, customs, and ecological conditions. For example, some indicators of sustainable land use developed by three Kalahari pastoral communities were similar among the communities while others were specific to one community and place (Fraser et al. 2006, Reed et al. 2008). While such specificity can be valuable, it also hinders comparison between sites and requires each community to undertake its own potentially time-consuming process of index creation. The city of Bristol in the UK avoids the former problem to a significant degree by utilizing both locally identified indicators as well as some developed by national and international organizations. In this way, it monitors local priorities while facilitating comparisons between cities and allowing the assessment of its progress toward national and EU goals (McMahon 2002).

Within the confines of their own communities the indexes studied incorporate adaptability in a variety of ways. For example, some communities studied by S. Lopez-Ridaura et al. monitored the adaptability of agricultural systems to new conditions by tracking factors such as the "adoption of new alternatives," "farmer's permanence within a system," or the "proportion of an area with an adopted technology" (Lopez-Ridaura et al. 2002: 142). The quality of life indicator project in Guernsey built adaptability into its methods as it enabled "island politicians, policy-users, and relevant stakeholders" to change the indicators developed by the government and an official research unit (Fraser et al. 2006: 120–122). Additionally, the very fact that they are working on a local level indicator rather than using a preexisting common index suggest that local index developers value adaptability to particular situations.

An ethical examination of sustainability indexes   167

Across the board, local sustainability indexes exemplify the principle of careful use insofar as they value decreased pollution and the exploitation of resources. Once again, we see innovation in these indexes, in this case, in the types of resources whose use is monitored and conserved. For example, several indicators of quality of life in Bristol monitor land and building use to encourage the repurposing of abandoned buildings, the redevelopment of brownfields, and mixed-use community planning. All of these endeavors promote careful use insofar as they emphasize the preservation of rural land, utilizing existing infrastructure and the preservation of rural land (McMahon 2002: 179–81). Indicators developed by Diana Gallego Carrera and Alexander Mack also align with careful use as they monitor expert estimates of the length of time known energy reserves will last at current rates and thus enable people to plan their resource use according to priorities about future generations (Gallego Carrera and Mack 2010: 1033).

The basic project of gathering information to define and monitor progress toward sustainability at a local level, especially when paired with local policy-making, enables people to recognize and take responsibility for their actions, whether regarding air pollution, water quality, recycling, or any other activity. Once again, some unique methods of monitoring responsibility were developed in local indexes. For example, community members in Bristol identified "the percent of Bristol business actively engaged in measurable environmental improvement," "the average carbon dioxide emissions from council homes," and "complaints of dog fouling" as important indicators of sustainability (McMahon 2002: 179–80). All of these indicators enable individuals, groups, and/or businesses to take responsibility for their actions by identifying their culpability and enabling them to change.

Yet local indexes can focus users' attention on their responsibility for local conditions so much that the effects of local activities on other places are overlooked. For instance, usually the only sign, if there is any, of the community being studied influencing the rest of the world is found in measurements of greenhouse gas emissions (e.g. McMahon 2002: 179). For this reason, the local indicator sets and indexes examined have a limited ability to foster responsibility and farsightedness.

Similarly, many local indexes are able to align with a part of the principle of justice *because* of their narrow focus which enables index developers to involve local stakeholders and policy-makers in the development of the indexes. These participatory methods enable people to interact with policy-makers, gain confidence and skills related to sustainability progress and policy-formation, and shape the policies that affect them, aspects of participatory justice. Distributional justice with respect to basic physical conditions for life are also monitored through indicators of food sufficiency (Praneetvatakul et al. 2001: 106), homelessness, water and home quality, education, life expectancy, long-term illnesses, and "the percent of homes and business with affordable energy services from renewable and efficient energy sources" (Fraser et al. 2006, McMahon 2002: 179). Some local indexes focus on conditions relevant for a particular, often disadvantaged, population such as "facilities for disabled motorists and pedestrians" (McMahon 2002), "aboriginal life expectancy at birth," or the proportion of women in government (Fraser et al. 2006). While most measures of equity emphasize social, as opposed to environmental conditions, Gallego Carrera et al. explicitly link the two when they track expert opinions of the "perception and fairness of risk distribution and benefits in neighboring communities" and the "subjectively expected health consequences of normal operation" of different energy sources (Gallego Carrera and Mack 2010: 1033). Their studies could be even more aligned with participatory justice if they directly surveyed locals rather than relying on expert opinion of what community members probably think. Despite this distance from actual community opinion, local sustainability indexes often methodologically align with justice through their involvement of local people in index formation. Additionally, they may include some measures of social equity and do so in a number of innovative ways when compared to national indexes. Yet local indexes still barely include methods of assessing issues of equity that are explicitly related to environmental benefits and burdens for humans.[3] As always, ethical limitations of the index arise in part from limited data and analytical methods. Consequently, the decision to abstain from tracking environmental equity can be a move to balance feasibility and idealism.

Not unexpectedly, local index projects vary considerably with respect to their exemplification of feasible idealism. Those that bridged this gap well tended to brainstorm ideals for indicators then narrow the set based on one or a combination of three factors: importance to locals, data availability or measurability, and verification by experts or scientific methods. More feasible indexes were less complex, relatively easy to measure and compile, and were constructed in conjunction with local policies (Fraser et al. 2006: 117–18, 122) as predicted by theories of index development discussed in Chapter 2. Local involvement in index construction was found to increase support for sustainability initiatives and the monitoring itself in Bristol, where people became more involved over time, increasing the feasibility of using the index (McMahon 2002).

In conclusion, local indexes, especially when developed with local input, enable the values, ecosystems, and community structures of their places to influence the indexes, leading to some novel indicators and emphasis on participatory justice while still including basic environmental and social measures (e.g. greenhouse gas emissions, life expectancy) found in many indexes. These indexes fall short ethically in two major ways. First, they may be so focused on a narrow vision of communities that they are not able to consider the impacts beyond the community and take responsibility for them. Second, the indexes barely begin to examine whether benefits and burdens of environmental and social conditions are distributed evenly throughout their communities. In this way, local sustainability indexes ethically mirror many national indexes.

5.7 Ethical strengths and weaknesses of sustainability indexes

Overall, sustainability indexes align with the broad principles of adequate assessment, farsightedness, adaptability, and careful use fairly well within the limits of available data. Admittedly, some indexes are more adaptable than others insofar as they apply to both developed and developing countries, a variety of situations, or are explicitly designed to be modified over time. Those such as the 2012 EPI, the Eurostat SDI, and some local indexes which are constructed with policies in mind are most feasible and enable responsible action even as they may press toward more ideal conditions. The greatest limitation of the indexes occurs with respect to the social dimension of sustainability; it is understudied in comparison with other dimensions. Indeed, its relative neglect is increasingly realized in the literature and tentative steps are being taken to overcome this lacuna (Assefa and Frostell 2007: 63, 66, Azar et al. 1996b: 91, Binder et al. 2010: 72, Burger et al. 2010, Pearsall 2010: 875). Yet the gap in the literature persists and contributes to the largest ethical limitation of the indexes: their limited ability to promote justice and the ramifications this has for their ability to embody the other interconnected principles. Admittedly, a number of indexes, including the three-dimensional index of SED, Prescott-Allen's HWI and EWI, the 2012 EPI, the Eurostat SDI, and some local indexes include some assessment of equity whether with respect to the access and affordability of high quality energy, equity of income among the population, political power among men and women, or disease burdens among the poor. Many local indexes also move toward participatory justice, in that they enable locals to contribute to the process of developing indexes. Yet, no index includes all of these elements and *none* monitors the distribution of environmental benefits or burdens within a nation or community.[4]

These results demonstrate that the broad principles, interacting with technical considerations, can provide significant feedback to the process of index development. Not only do they identify the ethical strengths of the indexes, but they also suggest a new line of research for sustainability index studies: examining whether and what lessons from measurements of environmental justice can be used to fill the most significant ethical gap in sustainability indexes, the subject of the following chapter.

Notes

1 Chapter 5 sections 4–7 are an expansion of Fredericks, S. E. (2012). Justice in Sustainability Indexes and Indicators. *International Journal of Sustainable Development & World Ecology*, 19(6), 490–99, reprinted by permission of Taylor & Francis (http://www.tandfonline.com).
2 Davidsdottir et al. recognize that many alternatives to GDP have been developed to include environmental resources and social goods, but do not use them since none has displaced GDP worldwide.
3 Though some intrinsic value theorists or biocentrists might like to see indicators focused on the biotic community aside from their relevance to humans, these

local indicators emphasize connections to humanity. This trend may be due in part to the fact that I chose indexes which stressed social aspects of sustainability, but I would be surprised if other indexes significantly emphasized nonhuman points of view given the state of index theory in general.
4 Some people would also find fault with the indexes because they do not emphasize the disparate benefits and burdens of environmental degradation across species, but since this assessment is based on a narrow, controversial specification of the principle of justice, and it has not yet been demonstrated that applying it would yield significantly different indicators, I hesitate to name it as an overall limitation of the indexes.

6 Environmental justice
A resource for sustainability indexes?[1]

In the preceding analysis, it was clear that the social dimension of sustainability is the location of many of the most significant ethical limitations across all types of sustainability indexes and that these indexes fall shortest with respect to justice. Given the interconnection between the ethical principles, limitations regarding justice often stemmed from or led to insufficiencies regarding other principles. For example, if index developers did not recognize that subpopulations could experience disproportionate benefits and burdens during their adequate assessment of the situation, they were unlikely to include a measure of justice in the index. Similarly, without such knowledge, they were unlikely to make indexes adaptable to the needs of various groups within a population. This neglect also hindered the indexes' ability to enable people to take responsibility for their actions, act farsightedly by looking to the impacts of actions around the globe and into the future, and implement policies that would realistically move toward sustainability. Thus, means of incorporating justice into the indexes are needed to move toward ethically robust sustainability indexes. Making this move requires a number of components: richer concepts of environmental justice, methods of monitoring progress toward them, data, and methods of incorporating environmental justice indicators into sustainability indexes. Environmental justice studies is a promising resource in the quest to address these issues because definitions of environmental justice align with and can enrich sustainability scholars' discussions of intra and intergenerational justice and because environmental justice scholars already monitor environmental justice. Yet, as we shall see, though environmental justice studies can significantly address limitations of sustainability indexes with respect to justice it cannot completely resolve them because of the nascent state of the field and its focus on the particular.

To explore the degrees to which environmental justice literature can enable sustainability indexes to more deeply align with the broad principles, this chapter proceeds in three sections. To justify the collaboration between sustainability studies and environmental justice, Section 6.1 revisits definitions of environmental justice introduced in Chapter 2, explores how they align with ideas of equity and justice in sustainability studies, and identifies how they can aid sustainability discourse. Section 6.2 describes methodological trends in monitoring environmental justice and identifies the most frequently monitored types of justice. Building on this information, Section 6.3 explores what methods of monitoring environmental justice can bring to sustainability indexes and the reasons they cannot sufficiently overcome the limits of sustainability indexes with respect to justice. In sum, this chapter will show that environmental justice studies can aid assessments of the situation in sustainability studies, reinforcing the need to include justice in sustainability indexes. They cannot, however, simply be incorporated into sustainability indexes because they often operate at different spatial scales, because environmental justice indexes tend to focus on narrowly defined issues, because comprehensive data sets are unavailable, and because robust methods of prioritizing and aggregating environmental justice issues have not yet been developed.

6.1 Definitions of environmental justice, their existing connections to and potential contributions to sustainability studies

Arising out of the first-hand experience of foul-tasting tap water, rashes, coughs, asthma, and cancers correlated with environmental degradation, the environmental justice movement began with local activists concerned about their families and community's health and well-being. The environmental justice movement grew as activists realized it was not enough to protest burdens in their own backyard; no one should bear such burdens. Scholars aided the movement by demonstrating that people of color and the poor are often less likely to benefit from clean environments and more likely to be burdened by environmental degradation caused to benefit others. This disproportionate distribution of environmental benefits and burdens has been deemed distributional environmental injustice. It is considered unjust because morally irrelevant characteristics (e.g. a person's race or ethnicity) influence one's exposure to environmental benefits and burdens. Participatory justice, the ability to meaningfully participate in decision-making about issues affecting one's environment, society, and self, is often defined as a separate type of justice though participatory injustices can certainly lead to or exacerbate distributive injustices while participatory justice can aid the quest for distributive justice. Recognizing that distributive and participatory injustices often arise from and perpetuate longstanding animosities between groups, restorative justice focuses on future relationships of those who perpetuate and suffer from environmental injustice, now or in the past. It aims to ensure that relationships between these groups are not only restored to conditions before the injustice occurred but improved to an unimpaired state, in which community relationships are strong and the potential for future injustices reduced. Both distributive and participatory justice have been significant concerns of the sustainability movement to date as its focus on intra and intergenerational justice often involves distributive and participatory concerns to identify problems, develop processes for identifying sustainability goals, and actually set the goals.

While there is already significant overlap between the content of environmental justice and sustainability studies, environmental justice terminology can bring precision to the sustainability movement. Environmental justice recognizes and names what is and names what is necessary to meet basic needs – goods and services, participation in processes, and social structures. Distributional and participatory justice also expand the conversation beyond discussions of the generations involved, as is often emphasized in the intergenerational justice discussions of sustainability studies. For example, insofar as environmental justice studies stress differences within a community, nation, or generation, they counterbalance the tendency of sustainability studies to focus on intergenerational issues. Environmental justice terminology can push sustainability studies to examine more detailed data rather than average characteristics of present populations and future possibilities. Thus, environmental justice's emphasis on the present may help raise support for sustainability initiatives, especially among people focused on daily quality of life.

Additionally, since past events and the situation of *all* in a community can affect material conditions and people's values of and for sustainability, the focus on past and present cultural priorities and situations in environmental justice studies could benefit sustainability studies. These insights will be particularly advantageous as sustainability indexes pay greater attention to the social dimension of sustainability, a move which will require deeper consideration of perceptions of cultural change and injustices as well as interpersonal and intercultural reconciliations.

Restorative justice studies will be particularly valuable in these endeavors given the impact of past and present environmental destruction and injustices on communities, nations, and international relations. The breakdown in international environmental policy discussions at Copenhagen was largely based on disagreement about who would be held responsible for decreasing carbon emissions to slow climate change: those who have already benefitted from carbon rich economies, those who are rapidly expanding their use of carbon as they develop, or some combination of both. This debate already significantly deals with past disparities in the use of environmental goods but tends to be forward-looking with respect to developmental opportunities and economic and environmental impacts of acting or failing to act. While considering future economic and environmental effects of climate change and new policies is certainly important, restorative environmental justice suggests that more is needed to resolve these issues. Countries must recognize their past and present responsibilities and culpability for climate change and harms to ecosystems and human communities in the past, present, and future. Restorative justice's emphasis on concepts of communal responsibility, guilt, and reconciliation could help reenvision future international relationships in which recognition and respect are prioritized so the distribution of contemporary and as yet unimagined future environmental benefits and burdens may be made more equitable while enhancing quality of life and ecosystem functioning. Since the theory and practice of restorative environmental justice are still quite new, more research is needed to clarify its concepts and develop its methods. Practices developed for dealing with apartheid and the Rwandan genocide after the fact may be helpful if reworked to apply to the often unintentional harms to people, biota, and ecosystems distant in space and time. By attending to these issues, sustainability initiatives will also be in closer alignment with ethical principles of responsibility, farsightedness, justice, and feasible idealism.

Given these realized and potential connections between environmental justice and sustainability it is clear that the terminology of environmental justice studies is a promising source for improving sustainability indexes' attention to justice as it is more precise and comprehensive. Yet methods developed for monitoring progress toward environmental justice and the data acquired through such methods can also aid sustainability initiatives.

6.2 Methods of monitoring environmental justice

Methods of monitoring environmental justice have been a significant part of environmental justice studies from the very beginning. Indeed, because many in the academic, political, and business worlds have not wanted to believe that environmental injustice occurs, environmental justice studies and action emphasized documenting distributional and participatory disparities and explaining why they are injustices. To date hundreds of studies have been conducted. Most focus on one narrowly defined issue (e.g. the location of toxic waste dumps), the issues facing a particular community (e.g. the South side of Chicago), or methodologies for assessing local injustices (e.g. involving locals in identifying and tracking injustices). This emphasis has enabled researchers to document injustices hidden in geographically larger studies and foster the participation of local people, a valued aspect of environmental justice. Through such research, scholars have demonstrated that environmental injustice is a multifaceted and widespread problem.

Methods of monitoring environmental justice can be classified into two broad categories: the first includes statistical studies and geographic information systems (GIS) analyses, both of which use quantitative data and typically monitor distributive issues (Corburn 2005, Goldstein 2005, Mennis 2002, Taquino et al. 2002). The second class of methods includes surveys and interviews which may monitor quantitative and qualitative aspects of environmental justice. This category monitors participation in decision-making, access to information, and perceptions of distributive justice. Examining these methods and their results will ground the study of what existing environmental justice studies can, and cannot yet, contribute to the development of sustainability indexes, the subject of the next section.

Early environmental justice studies aimed to ascertain whether or not environmental injustice existed and largely focused on the proximity of various racial and socioeconomic groups to toxic waste sites, assuming that proximity is correlated with risk and direct, experienced impacts. These studies tracked the number of toxic waste facilities or uncontrolled toxic waste sites within a certain radius of neighborhoods characterized by different demographic groups (Bullard 2000, Capek 1993, Commission for Racial Justice United Church of Christ 1987).

Later analyses revealed that observing environmental injustice could depend on the data used, size of study site, and method of determining risk (Anderton et al. 1995: 25–29, Bowen 2002: 7–9). For example, existing data about the addresses where toxins are located could facilitate tracking the number of hazards within a zip code or census tract. Yet this same data masks environmental risks to populations living just over the border into another zip code. Additionally, neither zip codes nor census tracts necessarily align with neighborhood demographics and census tracts do not map rural areas well.

In response, methods focusing on the distance to the hazard have been developed. Statistical methods determine the demographic characteristics of populations living within a certain radius of hazards. They enable researchers to determine the extent to which particular populations are disproportionately affected by the hazard in question (United Church of Christ Justice and Witness Ministries 2007: 39–45). GIS software has also been used to overcome some limitations of using zip codes and census tracts in hazard proximity studies. GIS analysts create maps overlaid with the demographic characteristics of communities and the severity of the environmental benefits and burdens they face (Evans and Marcynyszyn 2004, Maantay 2002, Mennis 2002). Though demographic data may be plotted using zip code or census tract data, hazards are precisely represented on the map, enabling users to observe the dangers within a certain radius of the population (Blodgett 2006). GIS studies can also represent multiple hazards on one map and help assess whether topographical features such as streams or wind patterns may aid or inhibit the spread of pollutants. Additionally, maps are often favored for communicating with the public as they can convey information about environmental injustice to laypeople unfamiliar with statistics.

Recognizing that the correlation of demographic groups and potential risks does not prove that these communities are harmed, some studies of environmental burdens move beyond demonstrating disparities in exposures or proximities to studying disparities in experienced effects. For example, studies have assessed whether individuals in certain communities have a higher than average risk of experiencing lead poisoning, developing cancer or lung disease, or giving birth prematurely due to environmental factors (Landrigan et al. 2010, National Research Council (US) Committee on Environmental Epidemiology 1991, Woodruff et al. 2003).

Even though environmental justice scholars are still developing means of monitoring environmental justice and determining best practices for identifying environmental injustice, evidence of environmental injustice continues to mount. Comparative research such as Paul Mohai and Bunyan Bryant's comparison of fifteen environmental justice studies indicate that "income and racial biases in the distribution of environmental hazards exist" and that racial and ethnic factors are often more highly correlated with proximity or exposure to environmental burdens than household income or wealth (Mohai and Bryant 1995: 20). Additionally, many studies chronicle the consequences of environmental injustice for the health and well-being of individuals and communities. For instance, Sacoby M. Wilson et al. maintain that the lack of infrastructures such as sanitation and water services can be a significant component of injustice (Wilson et al. 2008). In neighborhoods with uncontrolled toxic wastes, residents may have higher rates of asthma, kidney and lung damage, neurological diseases, cancer, brain hemorrhages, and birth defects (Bullard 1995: 18, Gibbs 1999: 28–30, National Research Council (US) Committee on Environmental Epidemiology 1991). Traditional local livelihoods can be disrupted when fishing and hunting grounds are destroyed through development or rendered unsafe by pollution. When fishing, hunting, and agriculture continue despite environmental degradation, invasive species, herbicides and pesticides, and health impacts of subsistence fishing in polluted waters more significantly impact people of color and the poor (Chavez 1993, Corburn 2005, Moses 1993, Norgaard 2007). Residents of poor and minority city neighborhoods also face significant health risks of urban heat island effects resulting from a relative lack of vegetation, high concentrations of buildings and pavement, few monetary resources to keep cool, and, increasingly, global warming (Harlan et al. 2008: 187). Environmental injustice can also stymie the potential for a community's future economic development if polluted local land discourages new businesses. Yet these consequences are not discrete events. Studies of the interaction between indigenous groups and relatively recent immigrants including traditional inhabitants and Australian miners in Papua New Guinea (Low and Gleeson 1998) and Native American opposition to a proposed road through, nuclear waste storage on, or military activities on sacred land illustrate how health, economic, cultural, and environmental concerns are interconnected in environmental justice cases (Checker 2002, Martin 2002, Shrader-Frechette 2002: 95–116).

Though environmental justice scholarship largely emerged in the United States, environmental justice scholars have been careful to note that environmental injustice is a global issue as they have documented various types of environmental injustice in communities around the world. For example Atanu Sarkar found greater effects of arsenic-tainted groundwater on the poor in West Bengal, likely because of their greater exposures through agricultural activities, malnutrition, and lack of health care (Sarkar 2009). Lim Weida found that climate change policy in Singapore fails to yield significant environmental benefits and reinforces structural inequalities regarding who bears the burden of environmental degradation (Weida 2009). The Stern Report and work of the IPCC have noted that global economic and environmental impacts of climate change will significantly impact the global South and those least economically and politically able to respond effectively (Intergovernmental Panel on Climate Change 2007: 48–52, Stern 2006: 59). Of course, these examples only scratch the surface of research on environmental justice which is now chronicled in many books and journals. The consistent theme in this growing literature is that burdens of environmental degradation fall disproportionately on people of color, minority ethnic groups, and the poor.

These results largely stem from traditional risk analyses, statistical proximity studies, and GIS studies. These methods all focus on types of injustices that can be relatively easily quantified: the number of superfund sites or other hazards in a region, asthma rates faced by particular demographic groups. This focus makes sense because mathematical methods require mathematical data. And yet many aspects of environmental justice are not easily quantifiable. It is much more difficult to study the direct effects of chemicals on human bodies and the correlation of such effects with ethnicity, race, and so on. If exposure rates are studied, scholars typically examine the effect of only one chemical at a time. While this approach eliminates the difficulties of studying multiple chemicals at once, it also significantly limits knowledge about the potential combinatory effects of exposure to multiple chemicals.

Even less studied are the social and psychological ramifications of environmental injustices. Deciding to limit the consumption of locally caught fish contaminated by heavy metals and PCBs may have significant implications if fishing and consuming local fish is economically necessary to obtain protein for one's family or at the heart of cultural and personal identity as in many Native American and Asian American populations (Corburn 2005: 79–110). Given the known interconnections between psychological stress and physiological stress, it is possible that disruptions of traditional cultural practices such as eating locally caught fish or performing a ritual at a sacred site may cause stress which may in turn impact the mental and physical health and well-being of individuals and communities. Yet it is not easy to quantify whether one can live in the style of one's ancestors, share rituals with one's family and neighbors, or preserve one's culture and the psychological health that may accompany these activities, especially using existing data.

To monitor such issues, environmental justice researchers tend to use the second large class of methods of monitoring environmental justice: surveys and interviews. These methods can track environmental economic and social insights from local ecological knowledge; assess people's access to information about environmental issues; register their ability to participate in decision-making; and monitor their perceptions of injustices and local sustainability (Checker 2007, Corburn 2005, Peeples and DeLuca 2006, Reed et al. 2006, Williams and Florez 2002). Some indicators of sustainability and quality of life have already used self-reported assessments of conditions in the community whether perceptions of noise (Eurostat 2012d), health (Fraser et al. 2006: 113), dog excrement, or crime (McMahon 2002: 179–80); or the participation in, use of, or confidence in governmental structures (Eurostat 2012d, Fraser et al. 2006: 118, 127, Gallego Carrera and Mack 2010: 1033). The development and use of these indicators shows that people are recognizing that perceptions of justice or sustainability are important aspects of achieving environmental justice and sustainability and that such data collection, while difficult, is feasible in at least some cases. Certainly, more work is needed to improve the financial feasibility and data collection methods of indicators of community-reported justice and sustainability. Such efforts are significant and should be encouraged to increase the knowledge they yield about justice and sustainability and because the ability to participate in indicator development and measurement helps ensure that indicators are meaningful to and implementable in local conditions while taking a step toward participatory justice.

Surveys and interviews are the most often used means of monitoring aspects of participatory justice. To monitor whether people have the ability to participate meaningfully in decision-making that effects them, some environmental justice advocates suggest monitoring 1) whether managers practice decision-making that takes into account the interests of all involved; 2) whether locals have been notified about upcoming decisions in ample time and through unbiased hearings; 3) whether they are consulted in the decision-making process; 4) whether compensation is given to those harmed by injustices; and 5) whether commitments to fight further injustice develop (Capek 1993: 8). Others identify targets for justice that ensure the empowerment of people such as "rebuilding of social capital (e.g. neighborhood cohesion), human capital (e.g. professional skills), physical capital (e.g. infrastructure), and natural capital (e.g. natural resources and living systems) to improve community health and restore neighborhood vitality" (Litt et al. 2002: 192). While these are all potentially valuable means of tracking participation, they all require additional specification to be transformed into indicators.

Surveys and interviews can also help ensure that local visions of environmental justice (or sustainability) are identified and progress toward them is monitored. For example, community members' own priorities, ascertained through pilot studies, may be used to structure questions about environmental justice. If the ability to bury the dead using traditional rituals in an ancient burial site is a priority for a community, survey questions could elicit information about the degree to which current environmental conditions help or hinder the fulfillment of this cultural priority. Surveys could gather information about multiple community-defined priorities to give community members and researchers a broader picture of the social impacts of environmental degradation. If, however, such complexity was deemed too expensive or time-consuming, surveys could ask a general question about the degree to which environmental conditions impacted community or individual quality of life.

Regardless of the number of questions asked, surveys and interviews enable researchers and communities to track what is important to locals, whether their goals change over time, and how their perceived ability to achieve their ideals has changed over time. These methods, especially if taken seriously in policy-making, not only help to monitor distributional environmental justice with respect to cultural, psychological, and aesthetic issues but also are a step toward participatory justice (Checker 2007, Corburn 2005, Peeples and DeLuca 2006, Williams and Florez 2002).

To date, the implementation of indicators of participatory justice lags behind the creation of distributional justice indicators, probably because theories of measuring participatory justice are still being developed and implementing them is temporally and monetarily expensive. Though monitoring participatory environmental justice is still relatively new and faces methodological and practical challenges, surveys and interviews are often the best means to assess various aspects of participatory justice and social, cultural, psychological, and cumulative effects of distributional injustices. Alternative means of monitoring participatory justice, which rely on existing data sets, for example, are often weakest with respect to the very group one wishes to examine. For instance, common means of monitoring civic engagement, a possible proxy for participatory justice, include rates of participation in community activities or clubs; newspaper readership rates; home ownership rates; or the proportion of minority participants in elected leadership roles, in the business community, or in the middle class. Yet, these standard measures of civic engagement may be biased against the very groups that environmental justice wants to study, such as speakers of nonmajority languages that may not have a local paper. Additionally, these measures do not ensure participatory justice; minority participation in elected leadership does not ensure that the average local citizen has a voice in decision-making or that the needs of minority groups with respect to environmental justice will be met. Thus, index developers focused on participatory justice will often need to rely on surveys and interviews instead of or in addition to preexisting data.

Surveys and interviews are increasingly used as methods of monitoring progress toward participatory justice and cumulative social, health, and psychological aspects of distributive justice. These tools, along with statistical methods of risk analysis and GIS studies, comprise the major means of monitoring distributive and participatory environmental justice. As restorative justice initiatives grow, similar tools will likely be used to monitor progress toward restorative justice (statistics will monitor whether conditions have improved materially and surveys will assess whether attitudes and relationships have changed). Though methods of monitoring all types of environmental justice are still being refined, results of environmental justice studies around the world indicate that environmental injustice is a widespread phenomenon.

6.3 Contributions of environmental justice studies to sustainability indexes

Given the overview of suggested and implemented means of monitoring environmental injustice, and the data which results from them as found in

Section 6.2, this section explores what environmental justice studies can contribute to sustainability studies, particularly index development, as well as factors which limit such contributions. Beyond the in-depth categorizations of types of environmental justice and explanations of injustices discussed in Sections 2.3.4 and 6.1, data about the widespread nature of environmental injustices as well as environmental justice indicators may contribute to sustainability studies. Theoretically, environmental justice indicators could be incorporated directly into sustainability indexes, to overcome their limitations with respect to justice. In most cases, however, significant developments in index construction and data collection are needed to incorporate environmental justice indicators into sustainability indexes.

The very act of documenting environmental injustices throughout the world is an important contribution to sustainability studies. If sustainability initiatives aim to ensure that the needs of all in the present and future can be met then information about the distance to this goal, as evidenced by current conditions, is valuable. Indeed, the evidence of persistent widespread environmental injustices as outlined in Section 6.2 suggests that it is likely that sustainability initiatives will also yield disproportionate benefits for different demographic groups unless specific steps are taken to avoid such outcomes. Yet, many indexes, influenced by prevailing economic theory, presume a "trickle-down" theory of justice with respect to sustainability as they assume that increasing the *average* quality of life or average access to environmental benefits, or reducing *average* exposure to environmental harms will improve the lot of *everyone* in the present and future. But why assume that improving average environmental conditions without explicit attention to the worst off would significantly benefit them when social systems already tend to favor others? For instance, grandfather clauses make trickle-down justice difficult. Such clauses allow existing facilities to maintain current levels of pollution though the legislation mandates stricter standards for new facilities. They are intended to ensure that existing businesses are not crippled by environmental laws. However, they also make it more difficult for people in the vicinity of the biggest polluters to improve their environmental conditions since owners of these facilities, which may be four to ten times as polluting as facilities that comply with new regulatory standards, often keep old facilities operating as long as possible to take advantage of the laxer environmental standards (Gorovitz Robertson 2008: 537–39). Thus, trickle-down theories of justice are insufficient bases for monitoring progress toward sustainability if sustainability implies that the needs of all in the present and future are met.

Limited data and the desire to use existing data sets to facilitate comparisons of past conditions may also encourage the study of average conditions rather than disparities in access to goods and services. Yet, the use of average data and a trickle-down vision of justice are seriously called into question by data about environmental injustices.

The ability to challenge conventional indexes of sustainability and conventional modes of assessing the situation is probably the most significant way environmental justice studies can contribute to the development of sustainability indexes. It suggests new directions for index theory and encourages the incorporation of environmental justice indicators into sustainability indexes, a move which can potentially overcome the main limit of sustainability indexes, the lack of attention to the just distribution of environmental burdens and participation in environmental decision-making.

While the combination of environmental justice indicators and sustainability indexes is a promising way to overcome limits of sustainability indexes, in practice, issues of scale, data availability, prioritization, and aggregation prevent such a simple solution. A systematic examination of both local and national indexes reveals these challenges.

Looking first to local indexes, we see that it is theoretically possible to incorporate local indicators of environmental justice directly into local sustainability indexes if both indexes apply to the same place at the same scale and if high quality longitudinal data about environmental justice measured at the same time interval as the components of the sustainability index is available. Under these conditions, environmental justice indicators would merely be some of the many indicators aggregated into the sustainability index. Yet local sustainability indexes are typically developed for a political entity such as a city or metropolitan area while measurements of environmental justice generally focus on particular neighborhoods. This spatial difference, along with the possibility of indicators being measured at different time intervals, hinders the aggregation of environmental justice indicators into sustainability indexes.

Even if sustainability indexes and environmental justice indicators are aligned in terms of scale, index developers will still face challenges of data availability and quality. Sustainability indexes are typically monitored in successive years to gain a sense of longitudinal progression toward sustainability. Yet most studies of environmental justice focus on conditions at one time so it will be necessary to develop ongoing data collection projects to incorporate environmental justice into sustainability indexes. Data collection may also be necessary because environmental injustice data for many areas is unavailable.

In the absence of ideal data, index developers often use a proxy. For example, when comprehensive data about environmental justice is not available for a whole city, data yielding a snapshot of one environmental justice issue in one neighborhood within a city could serve as a proxy for environmental justice within the city at large. Yet choosing a proxy is a difficult process because there is no guarantee that an available environmental justice indicator represents the physical situation faced in the whole study area, the diversity of people in the area, or local people's priorities. The comparative research needed to make such claims has simply not been done. On a local scale, it may be feasible to address such concerns through participatory processes in which locals, in consultation with relevant experts, could decide whether or not existing but narrow data should be used as a proxy for environmental justice, whether new surveys or measurements should generate environmental justice data, or whether it should be left out of the sustainability index. In any case, significant work is needed to ascertain how aspects of justice are to be prioritized in sustainability indexes.

In sum, there is potential to incorporate measures of environmental justice into local sustainability indexes, though such endeavors face significant challenges of scale, data availability, and prioritization. Unfortunately, these challenges are only exacerbated in national indexes of sustainability which also face the challenges of including particular local concerns in their national measurements.

As Section 6.2 demonstrates, the vast majority of environmental justice indexes operate at a local level, or are an issue-specific study of environmental justice across several locations. Few national studies of environmental justice have been conducted. Thus, data limitations inhibit the incorporation of any existing environmental justice indicator into a national sustainability index.

Related to this problem is the fact that neither index developers nor activists have significantly studied how to incorporate local environmental justice indicators into national assessments of environmental justice or indeed, methods of incorporating local indicators into national indicators in general. For example, there is no national environmental justice index comprised of many environmental justice indicators nor is there a framework or method to aggregate such indicators into an index. Indeed, the very concept of an overarching or standardized measure of environmental justice across different places and communities may be perceived as opposing the focus on local situations and values often taken as an act of justice in and of itself. Additionally, as of yet, there is very little, if any, discussion about which environmental justice factors are most environmentally, culturally, materially, and psychologically significant, and no guidelines for prioritizing them have emerged other than a vague focus on meeting basic needs, say as defined by the HDI or MDGs. These priorities, however, focus on environmental benefits, not eliminating environmental burdens. They also focus on national averages and assume that benefits will trickle-down to all to the degree that they cannot register many environmental injustices. Determining methods of representing the particular in overarching or national assessments will be key to identifying environmental justice indicators for scales other than the very local.

Further complicating this process is the fact that environmental injustice studies have not evenly covered the United States or the world. Many aspects of environmental justice have been studied in some places; in others no studies have been conducted. Without consciously attending to the question of how to prioritize environmental justice indicators it is likely that those most studied will dominate whether or not these are the most significant for local people or for the nation as a whole. Using representative issues, communities, or groups of people as proxies for national assessments of environmental injustice will, of course, raise important questions about which groups should be the representatives, which types of justice should be focused upon, who would make such decisions and how such decisions would be made while respecting the needs, values, and situations of local people. Additionally, users of such proxies should acknowledge that prioritizing them may reinforce the tendency to focus on environmental indicators rather than on social, cultural, and psychological issues. While it will not be easy to come to consensus about such issues, regional or national assessments will be key to identifying environmental justice indicators for scales other than the very local. Some way of monitoring environmental justice is needed or sustainability indexes will continue to be limited with respect to the broad principle of justice because they will continue to rely on trickle-down methods of monitoring justice in the indexes.

In conclusion, the nascent state of environmental justice indicator methodology and data collection generally prevents the direct incorporation of environmental justice indicators into sustainability indexes at this time. Thus, environmental justice indicators are not a simple solution to resolve the ethical limitations of sustainability indexes with respect to justice, alone or in relation to the other ethical principles. Yet the potential to overcome this lacuna in sustainability indexes is significant since definitions and initial data from environmental justice studies have and can contribute so much to sustainability studies and methods of monitoring environmental justice, though still a relatively new enterprise, are rapidly developing. Additional research on issues of the scale, aggregation, and prioritization of indicators and indexes may yield significant dividends for sustainability indexes with respect to justice. To begin to conceptualize how this work could enable sustainability indexes to better align with the normative goals of the sustainability movement, and thus track and promote sustainability, the next chapter will explore and develop new methods of prioritizing and aggregating local indicators in national sustainability indexes.

Note

1 Chapter 6, sections 2 and 3 on monitoring environmental justice are a revised and expanded version of Fredericks, S. E. (2011). Monitoring Environmental Justice. *Environmental Justice*, 4(1), 63–69. This material is reprinted with permission from ENVIRONMENTAL JUSTICE, Volume 4, Number 1, published by Mary Ann Liebert, Inc., New Rochelle, NY.

7 Aggregating local indicators

Issues of scale and data availability, prioritization, and aggregation are the main factors hindering the direct incorporation of environmental justice indicators into sustainability indexes. Yet they also challenge efforts to incorporate local sustainability knowledge, values, and indicators in general into regional or national sustainability indexes. While local sustainability efforts have been encouraged at least since the first Rio conference in 1992 (Holden 2011: 313), few methods exist to aggregate local knowledge and values into national indexes while preserving something of their specificity (Mascarenhas et al. 2010: 647–48). Those that do exist are far from ideal. Yet coordinating sustainability efforts with national indexes is necessary to ensure that movements toward sustainability in one place are not offset by setbacks elsewhere, that best practices for sustainability policies can be identified, and that progress toward policy targets is made. Consequently, new methods of aggregating local measures of justice and other aspects of sustainability are needed to enhance sustainability indexes. This chapter identifies two such methods. Section 7.1 extends the work of Maria Luisa Paracchini et al. to yield analytical modes of aggregating local data into regional or national indexes. Section 7.2 notes how maps of environmental injustice can communicate the scope of environmental injustice, trends in the data, and data limitations to lay people. Finally, Section 7.3 explores how these methods, singly and in partnership, take steps toward overcoming challenges of scale, aggregation, prioritization, and data availability which can hinder efforts to integrate local sustainability and justice assessments into national indexes.

7.1 Numerical aggregation methods

While the literature is beginning to recognize the challenges of integrating the findings of local (or any small-scale) indicators into national (or any large-scale) indexes, these subjects are still understudied topics (Mascarenhas et al. 2010: 647–48). All existing methods of integrating indicators applicable at multiple scales face significant limitations. Generally, attempts to link local and national indexes depend on a small set of indicators used at every scale to facilitate aggregation and comparability between scales and places (Custance

2002, Mascarenhas et al. 2010: 649, Paracchini et al. 2008, Paracchini et al. 2011, Pérez-Soba et al. 2008). The use of consistent indicators can aid comparison between places and enable local data to be represented in regional and national indexes, but their use means that locally specific aspects of sustainability do not become indicators. This prevents place-specific concerns from being registered by national indexes. Furthermore, preexisting lists of indicators, upon which such indexes are based, typically emphasize environmental over social issues. Thus, cultural and aesthetic values as well as disparities among demographic groups will be neglected by such indexes.

This trend is particularly problematic if one is trying to incorporate environmental justice concerns into sustainability indexes since the very features one wants to incorporate into the national-level index – the particular group experiencing the injustice, the type of injustice suffered, and perceptions of injustice – are the very thing that indexes developed with this method cannot register. On the other hand, local index initiatives that facilitate monitoring sustainability priorities particular to a community are not able to be easily incorporated into national indexes. For example, a sustainability indicator project in Bristol, UK identifies and monitors sustainability indicators at many levels from the most local to the EU scale, but the most local indicators are not represented in regional, national, or EU level indexes unless they are a part of previously existing standardized indicator sets (McMahon 2002). Thus priorities and conditions specific to local communities and ecosystems do not sufficiently influence national indexes.

There are, however, groups of scholars, namely Maria Luisa Paracchini et al. and Ioannis Spilanis et al., whose work takes steps toward enabling local priorities to directly influence national indexes. Spilanis et al. allow categories of indicators to be specified by different indicators in different places[1] and Paracchini et al. note that different communities could use different methods of aggregating indicators (Paracchini et al. 2008, Paracchini et al. 2011, Spilanis et al. 2009). An extension of these frameworks, particularly relying on the work of Paracchini et al., will enable index developers to ensure that local priorities can influence national indexes, addressing limitations of sustainability indexes with respect to justice and social aspects of sustainability in general.

Paracchini et al. develop an index framework to monitor the sustainability of land use that can be implemented at either a regional level or a pan-European level. Forming the basis of the framework are nine land use functions identified by Pérez-Soba et al., including health and recreation, cultural functions of land use, land based production, infrastructure, "provision of abiotic reserves," support of biotic resources, and "maintenance of ecosystem processes" (Paracchini et al. 2011: 74–75, Pérez-Soba et al. 2008: 380–89). Paracchini et al. propose that changes in these nine land use functions be tracked by thirty indicators selected by an interdisciplinary team of experts from a list of forty indicators relevant to EU policy (Paracchini et al. 2011: 72, Pérez-Soba et al. 2008: 391). Indicators from the list were selected for their "direct or indirect causal links to the [land use functions]" and relevance at regional levels. Care was taken to ensure that the set integrated environmental, economic, and social issues without redundancy (Pérez-Soba et al. 2008: 391). Thus the indicators are supposed to span the whole spectrum of land use functions in ways that can directly monitor progress toward policy goals.

They propose a combination of three weighting factors to aggregate the land use function indicators. The first equally weights each land use function. The second registers the strength of connections between individual indicators and land use functions, as determined by a survey of experts, to register whether and to what degree indicators influence multiple land use functions. The third "reflects the importance of each indicator at a regional level," a determination that was made by expert analysis here but, as the authors themselves note, could also be influenced by stakeholder input (Paracchini et al. 2008: 369, Paracchini et al. 2011: 75–76, Pérez-Soba et al. 2008: 391). Paracchini et al. intend indexes used to assess the sustainability of the European Union to use the first two weighting factors. This method enables different types of land use to be weighted equally while avoiding double counting of indicators which influence multiple land use functions. They propose that regional indexes also use the third weighting factor to ensure that regional priorities dominate the index. For example, they suggest that an indicator of vulnerability to forest fires be given a weight of zero in regions with no such risk (Paracchini et al. 2011: 76). Thus, the land use functions incorporated into the index, and subsequently the sustainability index itself, may vary between regions. This variability enables regional indexes to focus on the aspects of sustainability that pertain to them even as the same framework guides the development of EU indexes (Pérez-Soba et al. 2008: 396).

The variability between regional representations of any one land use function is constrained in Paracchini et al.'s framework because they only enable regions to draw on a narrow, standardized, preexisting set indicators (Paracchini et al. 2011: 72). The indexes made from this method will be unable to monitor local or regional priorities that are specific to that location, the key feature for resolving many issues of prioritization, aggregation, and data availability in index development. To capitalize on this possibility, Paracchini et al.'s framework can be generalized following Spilanis et al. to be open to the possibility of locally defined sustainability indicators in addition to those already existing in common sets (Paracchini et al. 2008, Paracchini et al. 2011, Spilanis et al. 2009).

Under this model, community index developers would use indicators from standardized lists to facilitate comparison between communities and aggregating local indicators. In addition to these standardized indicators, locals could also identify indicators that represent locally valued aspects of sustainability. Community defined indicators could be completely different between places but still impact national indexes if they were aggregated into a "community defined" subindex of the national index. For instance, one community might monitor the degradation of a sacred river while another tracks how culturally important recreational activities are limited due to air pollution. Since such indicators are likely to be monitored using different units, and may even be a mixture of quantitative and qualitative units, index developers would need to normalize the data to a standardized scale in order to combine them in one subindex. With such a normalization process, multiple different community indicators for each individual community could be tracked, an important possibility if communities identified several locally specific priorities and index users wanted to know whether some communities were making more or less progress toward sustainability. If index users wanted to emphasize types of sustainability more than the comparative success of communities, several general categories of place-specific indicators such as perceptions of culturally relevant environmental degradation, disparities in environmentally related quality of life, disparities in proximity to hazards, or disparities in participatory decision-making could be included as distinct subindexes in an index. Then rather than aggregating all of the indictors from each community before combining them, indicators could first be combined according to topic regardless of community. This method enables users to assess whether progress was being made with respect to specific categories of sustainability more than others. Using any of these methods, local priorities could influence national indicators even if they monitored different aspects of sustainability.

Admittedly, enabling so much diversity within a sustainability index is an ambitious goal. Significant work to identify and develop local indicators of sustainability and measure data for them is needed. It is not feasible for this process to happen in every neighborhood or town. Thus, it is likely that representatives in a nation will be used. This simplification process is, of course, nothing new. Data from selected locations are often used as proxies for countrywide trends in environmental and sustainability assessments. For example, national air pollution data is often based on conditions in large cities where data is abundant and many people are affected. Following this example, and focusing on environmental justice, sustainability indexes could focus on environmental justice measurements from locations where people have most often protested injustices or where disparities have been formally documented. Such selection methods will probably skew the data toward the more problematic cases, but that is desired if one prioritizes the most vulnerable, a key assumption of environmental justice. If one valued participatory justice, locations should be chosen in a representative fashion, ideally through a participatory process. If the locations selected were found to be physically or socially inadequate, additional collection points for local data could be added to the index over time. With this caveat of adaptability, using data from well-studied environmental justice sites is a feasible way to establish baseline data and kick start the process of assessing environmental justice within sustainability indexes.

Certainly, these numeric methods of aggregating environmental justice indicators from many locations into national sustainability indexes will need to be specified according to the goals of the indexes, physical and societal conditions, and the availability of data. Significant attention must be also paid to the feasibility of these suggestions as their complexity could impede their implementability. Yet, since sustainability, including justice, is a significant goal of humanity today and people expect to be able to monitor progress toward it, then some method of incorporating justice into sustainability is required. Furthermore, since a foundational assumption of sustainability discourse is that it can manifest in different ways depending on not only physical circumstances but also normative values, enabling large-scale indexes to capture local variations in justice and other social concerns will be key to developing sustainability indexes that take justice seriously. Monitoring programs for air and water emissions, inconceivable several decades ago, are now widespread. Thus it is not inconceivable that, if prioritized, methods of monitoring environmental justice may follow a similar trajectory.

The methods outlined above, while in need of refinement in order to be implemented, are steps toward accomplishing these goals.

7.2 Map-based aggregation methods

While promising numerical methods of aggregating local sustainability priorities, including measures of environmental justice, into large-scale sustainability indexes exist, these methods leave something to be desired. Insofar as they focus on available data, they draw attention away from places and issues for which little data exists. Additionally, by nature, they tend toward aggregation which can draw attention away from the distinctiveness of particular environmental justice situations or of local priorities for sustainability in general. Furthermore, their numerical methods may be difficult for lay people to understand. Maps of existing environmental justice data can overcome at least some of these limitations and therefore complement numerical indexes. While all sorts of locally defined sustainability issues could be represented on maps, this section will focus on environmental justice because of the relative neglect of environmental justice in large-scale sustainability indexes.

Maps representing existing environmental justice data will help ensure that environmental justice data is not taken to be more certain than it is, that policy-makers and the public can readily understand basic ideas about environmental justice, that local decision-makers have tools to help inform their choices about the development and environmental cleanup endeavors, and that areas which are ripe for future research can be identified.

At minimum, maps to represent environmental justice should meet six criteria. First, they need to be capable of registering distributive disparities between populations, rates of participation in decision-making, or disparities between access to conditions necessary for such participation (e.g. knowledge and the formal and informal structures which enable participation). The maps also need to represent the variety of environmental injustices studied (e.g. proximity to toxic waste, urban heat island effects, etc.) and the various groups (e.g. economic, racial, ethnic and maybe age and gender) which experience these injustices. Third, since types of environmental injustices may evolve over time and may affect different groups in different places at various times as new burdens are created, new laws made and enforced, and populations move, the maps should also be able to capture temporal changes. Fourth, the maps should have a national scope to illustrate trends in national environmental justice, the scale most neglected to date. Yet, if they are to complement the averaging that may occur in numerical aggregations the maps should also preserve the local flavor of data. Finally, since studies of environmental injustices not only document the fact of injustice but also monitor its severity, maps must also represent both of these aspects. Let us explore how each of these conditions can be represented on maps.

The severity of environmental injustice is usually presented in one of three ways. First is the disparate chance of exposure to environmental harms. For example, one demographic group may be a certain number of times as likely as another to live within a specified distance of toxic waste facilities; be so many times as likely to suffer from asthma, lead poisoning, or other environmental diseases; or be so much less likely to be meaningfully involved in environmental decision-making. The second way to track the severity of environmental injustice includes data about differences in degree of harm which use a measurement other than percentage to compare two groups. For instance, one may monitor the temperature differential between a neighborhood of color and a white neighborhood during a heat wave (Harlan et al. 2008). Here, the focus is on what is experienced differently (heat) rather than different risk rates. Finally, severity of environmental harms may be monitored by perceived disparities in risk, as when Mexican Americans and Caucasian Americans were found to have different perceptions of "environmental inequity" or when socioeconomic status was linked to perceptions of water-quality risks (Williams and Florez 2002).

Since the magnitude of these findings is as important as the fact that disparities exist, evidence of environmental injustice will not be represented well by a simple symbol, say a circle or an x, to indicate where on the map environmental injustice occurs. To indicate the magnitude of the injustice, I suggest that cartographers use "mountain peaks" centered on the location of the injustice with the height of the peak rising with the severity of injustice. Maps focused on percentage-based measurements of risk will be easier to make because the severity of multiple types of risk will already be represented on a standardized scale. For other sorts of data, normalization will be required to represent the severity of environmental injustice chronicled by various studies on a single map. Because scales for normalization will depend on priorities of index developers and probably stakeholders, some may opt to focus on data that are already normalized to obtain initial maps of environmental justice to supplement sustainability indexes.

While static maps could certainly aid the assessment of national trends in environmental justice, interactive, computer-based maps may be most informative for users. With such maps, users could select a certain injustice – say disparity in proximity to toxic waste facilities or differences in urban heat island effects – and see a map of the United States with all such data for a variety of groups. Alternatively, one could track the various disparities experienced by a particular group, say Native Americans, African Americans, or Hispanics, or the comparative disparities between any two groups. Different colors, shades of color, or textures could indicate different issues or groups studied. So, for example, blue may indicate disparities in proximity to toxic waste, diagonal lines may indicate disparities between Latinos and whites. Thus a tall, diagonally striped blue mountain at a place on the map would indicate a severe disparity with respect to Latinos living near toxic waste facilities. If enough color or texture was available, and normalization could occur, all tracked injustices across all groups could be monitored on one map.

Additional information about an injustice spike could be accessed by a pop up window visible when one clicks on or hovers the cursor over the spike. Such boxes could include study parameters such as the temporal and spatial boundaries of the study, questions used to assess environmental justice in surveys, and statistical methods used to calculate the severity of injustices. Pop up windows could also include citations of the published information or web links to the relevant articles or data sets.

Maps could also be constructed to compare conditions between time periods. Either static maps focused on different time periods or an animation which moves to a new year or decade's data every ten seconds or so could represent such changes over time. If data was detailed enough, and policies were represented on the map, say by outlining affected areas, a viewer could track effects of policy changes on injustices.

Of course, since some locations have been the subject of many environmental justice studies, national maps tracking measured injustices may appear to indicate that Houston, for example, experiences greater rates of environmental injustice than other places, say North Dakota, only because there is an abundance of data about the former and little to no data about the latter. To guard against such hasty misconceptions, I suggest that areas which lack environmental justice data be represented by a transparent cloudiness that, while evident, does not obscure geographic features such as state boundaries, cities, and bodies of water. Such a cloud would stand for the murkiness that results when information is unavailable and the likelihood that some haze of injustice occurs in that location even if one cannot map it with existing data.[2]

With such features, environmental justice maps could quickly and easily inform policy-makers and citizens about national trends in environmental injustices while preserving local specificity. If the maps included the fog of missing information they would also highlight places, injustices, and demographic groups which are understudied and may be fruitful avenues for new research. Such maps could aid local decision-makers in the identification of locations for new potential hazards, in the allocation of clean-up funds, and in initiatives to increase community participation in environmental decision-making.

Admittedly, EJView, an interactive mapping tool developed by the EPA, already yields the ability to identify potential local hazards, areas where injustices may occur, and aid local decision-makers in their decisions about when to take action against injustices (Environmental Protection Agency, Robertson 2008: 538–39). This interactive map system enables community organizers, planners, and others to plot sites regulated by the EPA including sites of hazardous wastes, air emissions, water discharges, and toxic releases as well as superfund and brownfield sites; demographic features including population density, per capita income, educational levels, age, minority status; health statistics and locations with poor health services; locations of particular interest to environmental justice studies such as schools and hospitals; and geographic features such as water sources. Users can select whatever region in the United States they wish and generate maps with their preferred characteristics from the list above. Often, data is available at multiple scales – blocks, tracts, and counties – so that the program can benefit people interested in different scales and so users can assess whether injustices only appear at certain levels. EJView is intended to enable communities to document potential problems to aid community education, outreach, and grant-writing endeavors (Environmental Protection Agency). It could also be used by local zoning and planning commissions as new facilities are proposed and laws contemplated.

EJView, however, does not enable national assessments of environmental justice because it is built to focus on local assessments of proximity to risk which require fine-grained rather than regional or national scales. Data on toxic releases, superfund sites, and so on simply does not show up accurately at larger scales on EJView because many sites are left off the map or grouped into one rather inaccurately located symbol at large scales. Additionally, the map does not allow a user to readily assess the severity of potential and actual environmental justice issues because it uses a basic symbol for superfund sites and places monitored for air quality or water emissions. Data about the sites' chemical emissions per year is available only if one clicks on the site (Environmental Protection Agency). While this data would certainly be helpful for local organizations who desire information about risks they face, EJView cannot represent the severity of risk on the map itself aside from tracking numerous hazards simultaneously. The severity data linked to the map is way too detailed to provide an overall assessment of environmental justice to accompany a national sustainability index.

EJView also falls short of the needs of a national environmental justice assessment because of the limitations of its capacities for representing demographic data and the types of risks it monitors. For instance, while EJView includes many demographic factors it does not enable assessments of comparative risks between different minority groups (Environmental Protection Agency), a common question in environmental justice research, because it lumps them all together. Additionally, EJView only displays one demographic feature, the most recently selected, at a time, making it impossible to simultaneously track multiple demographic characteristics, such as poverty and minority status (Environmental Protection Agency). Finally, EJView focuses on tracking disparities from toxins, which makes sense given the available data and focus of policies, but is not helpful for tracking community-specific or community-defined justice issues such as degradation of a sacred hill or traditional livelihood, participatory justice, or newly identified environmental justice issues such as disparities in experienced heat island effects.

In sum, while EJView can be quite useful for monitoring proximity to risks at a local level, its local focus and inability to address a variety of social issues means that it is not a sufficient tool to enable national assessments of sustainability as are needed to supplement sustainability indexes. Another set of maps, which represent the severity of measured environmental injustice in key places throughout a nation as described above, is necessary to meet these needs. Of course, limitations of data availability and quality can inhibit the utility of maps of environmental justice as can the potential for information overload for users. Though these challenges exist, the access to larger amounts of data over time and careful map creation that does not overwhelm the user with information can limit the effect of these challenges. Additionally, the promise of map-based efforts at tracking environmental justice on a national scale while enabling local priorities to manifest themselves is significant and justifies the effort needed to deal with their challenges. Since this mapping method can also highlight areas for future research, facilitate communication with the public, and aid local decision-making, it can confer multiple advantages to sustainability studies.

7.3 Advantages of numerical and map-based aggregation methods

Given the complexity and dynamism of the technical and normative priorities involved in sustainability, the challenges of scales for data, data availability, and the prioritization of indicators will never be fully resolved. Yet the frameworks for numerical indexes and maps of sustainability, including justice, considered singly and together, take significant steps toward addressing these issues so that local priorities can impact wide-scale sustainability indexes.

Numerical aggregation frameworks enable many issues important to particular communities to have some effect on national assessments of sustainability because they do not merely look at average data or issues that are important to all places. Therefore they enable a deeper representation of the local in the national, a step toward resolving problems of scale in sustainability indexes. The numerical aggregation framework also moves toward resolving the aggregation challenges in that it outlines several ways that diverse local indicators can be aggregated. I refrain from advocating one of these methods over the other because this choice will depend on the goals of the index. Similarly, allowing variability in the indicators that represent a component of sustainability such as justice in a sustainability index eases the challenge of prioritization since universal decisions about prioritizing conceptions of justice are not needed. Community members and experts can decide which indicators should represent a particular place. Of course, these methods do not completely resolve the prioritization problem because determining how community priorities and expert analyses should work together to select indicators for a place will not be easy. Additionally, selecting representative areas to serve as proxies, a potentially contentious aspect of prioritization, must be determined on a case-by-case basis. This decision will be influenced by index purposes, stakeholder priorities, and data availability. Insofar as communities propose indicators of environmental justice which are greatly valued in their community, there will be more political and public motivation to collect data than if the indicators were chosen by an outside group and or were not seen as relevant to the community in question. Thus, the numerical aggregation framework also takes a step toward overcoming the challenge of data availability.

In sum, the numerical index framework does not completely resolve the challenges of data availability, prioritization, aggregation, or scale but it does move toward addressing each one. Similarly, the map-based methods for monitoring environmental justice take steps toward addressing the challenges. User-defined maps enable different issues to be prioritized as desired, making such priorities more transparent. They will enable the simultaneous consideration of multiple types of local data across many locations as they can represent national or regional trends at a time or over time. Indeed, the particularities of index locations are preserved in the maps more than in the numerical aggregations for indexes though they require additional interpretations. Finally, if maps include a cloud of uncertainty they will highlight limitations in data and thus help identify areas where more research is needed, a way to move toward overcoming the challenges of limited data.

Both numerical and map-based methods could have a significant role to play in the deeper incorporation of justice concerns into sustainability indexes and decision-making. Realistically, it is likely that numerical methods will be preferred to simplify the process and because maps separated from existing indexes could easily be overlooked in favor of numerical results which are often understood to be more clear, objective, and true. Maps could provide a helpful supplement to numerical indexes by graphically racking and representing a host of place-specific sustainability issues in national indexes and representing disaggregated data that may be more comprehensible to lay people and enable the limits of data to be highlighted in ways that help preclude the ascription of absolute certainty to index results.

While the examples in this chapter focused on justice since its neglect is the most consistent of the ethical limitations of the sustainability indexes, the numerical aggregation and map-making methods outlined in this chapter could also be used to track and represent a host of place-specific sustainability issues in national indexes related to humans or other entities if desired by index users. The social dimension of sustainability would benefit most from such moves since it is so understudied and because cultural priorities about ecosystems can vary significantly from place to place, but environmental and economic dimensions could also benefit from the ability to vary indicators and represent various sorts of data on maps.

Even if map-based methods of assessing national trends in environmental justice or social sustainability do not take off, the ability to use locally specific indicators to monitor environmental justice enables national assessments to register the unique conditions and concerns of particular communities and ecosystems. Such a move can catalyze research and enable sustainability indexes to more deeply align with the broad principles of sustainability and conditions of injustice in the world, necessary requirements if the indexes are to better monitor progress toward both technical and ethical aspects of sustainability.

Notes

1 Prescott-Allen's species and genes index is the only other one I have seen with this feature. It was, however, buried in *The Wellbeing of Nations* (Prescott-Allen 2001).
2 Katy Börner's lecture "Science Maps as Visual Interfaces to Digital Libraries" inspired this idea. In that lecture, she argued that one can learn a great deal about how to represent new information, in her case, relationships between the sciences, from old cartographic practices. One of her many examples involved several hundred year old maps on which mapmakers simply stopped coastlines when the coast was unknown to them. Börner encouraged contemporary cartographers to follow this old practice and explicitly indicate when they lacked information (Börner 2011).

8 Ethics in sustainability indexes

Strategies for overcoming ethical limitations of sustainability indexes with respect to justice are crucial if people desire 1) sustainability as exemplified in the sustainability movement with its presumptions of justice and 2) a way to monitor and encourage progress toward it. Yet even if the data and methods to incorporate justice into sustainability indexes were available, they would not be sufficient. Sustainability involves other ethical principles and normative priorities can arise or change over time. Furthermore, general methods of integrating ethics into index development are needed to fix the underlying cause of the ethical limitations of indexes: the bifurcation of ethical and technical aspects of sustainability.

As Chapter 1 demonstrated, ethical and technical considerations are intimately intertwined in the articulation of sustainability definitions and goals as well as in the basic assertions that contemporary life in general or particular actions or mindsets are unsustainable. These normative priorities arise from the commitments of a variety of worldviews, whether ideas of a loving creator God coupled with ideas from modern science as favored by Nash, a focus on a God who is one and bestows responsibility on humans as described in Islamic law as articulated by Llewellyn, or in intrinsic value of present and future entities as espoused by the deep ecologists. Though these foundations for sustainability ethics may vary considerably, as can the details of the way ethical principles are specified by particular worldviews, key features of these ethical systems are represented by the broad principles of adequate assessment of the situation, adaptability, farsightedness, responsibility, justice, careful use, and feasible idealism. The ethical concerns infuse the sustainability movement.

Despite the implicit influence of technical and normative concerns on sustainability discourse and sustainable index development, guidelines for index development generally focus on technical considerations. Thus, Sections 3.5 and 4.4 outlined a method of incorporating ethics into index development, particularly the use of broad principles to evaluate existing indexes. Chapter 5 undertook such an assessment, demonstrating that existing indexes do not live up to the principle of justice because they can not track distributional differences in environmental benefits and burdens within a nation or community and presume a trickle-down model of justice. Indeed, ethical limitations of the indexes most often arise when data sets or analytical tools for monitoring aspects of sustainability are unavailable. The analysis also revealed that social aspects of sustainability are given less attention in sustainability indexes than economic or environmental dimensions; that individual indexes tend to be geared toward industrialized or developing countries and average populations and thus are not as adaptable to different situations as they might be; and that indexes tend to promote responsibility and careful use.

Aside from specific insights of broad principles alone or in combination, the general philosophical theory of vagueness can also contribute to index development when numerous different local conditions need to be aggregated into a national index. Traditional index theory supposes strict consistency of indicators between places to facilitate comparability. The broad theory of comparison, on the other hand, suggests preserving the unique features of individual cultures, worldviews, or places within broad categories of indicators. Therefore, indicators specific to cultures or ecosystems can influence national indexes, enabling an overall assessment of progress toward sustainability. Indicator variability is also supported by particular broad principles. For example, adaptability and assessment of the situation emphasize the need for indicators to align with local conditions. Responsibility also often favors such variability to ensure that locals and national populations could assess their contributions to and unique ways to move toward sustainability. Thus, the theory of vagueness and the principles developed through this theory serve as a philosophical bridge between the cultural specificity valued in environmental justice studies and the desire for aggregation in order to monitor general trends that is the impetus of index studies, especially on a regional, national, or continental level.

The ethical analysis of existing indexes and the implications of the theory of vagueness can suggest new subjects and methods of research to better align with sustainability goals. So, one may wonder if it is necessary to weave ethics into the process of index development, or if an assessment after indexes are implemented is sufficient. After all, a reactionary approach, in which normative priorities are explicitly considered after technical assessments are entirely or nearly complete, is common in many fields including bioethics, engineering, and environmental ethics. Yet explicitly considering ethics only after indexes are constructed and data collection begins is less likely to significantly impact the index for a host of reasons: Indexes are often used to track system changes over time through calculations of the index at regular intervals. Such comparisons are easiest if an index is composed of the same indicators and calculated in the same way over time. Index consistency can also facilitate the education of lay people and policy-makers about the index. Once an index is implemented, however, its momentum may make it difficult to change. Improvements to the index, whether suggested through ethical or technical analyses, may well be ignored in favor of the status quo.

Additionally, late-stage revisions for ethical reasons will generally be seen as supplemental rather than necessary information – the first to be cut if budgets or time are tight. Even if people accept the need to revise an index, aligning it with ethics after the fact may be an arduous process. While an extra indicator may be all that is necessary to fix the problem, an adequate solution may necessitate a radical rethinking of the whole index, as when indicator variability is introduced. Such revisions will almost certainly complicate the index development process beyond what it would be if explicit ethical analysis was involved from the beginning. Finally, the myth of index development as a technical process is perpetuated if ethical assessments of indexes only occur after they are created – ethics will be too easily seen as external to and separated from indexes themselves. When ethics is relegated to a minor consideration, it can be brushed aside or forgotten. Then sustainability initiatives and sustainability indexes will be vulnerable to charges of irrelevance or of pushing society away from its goals because they do not align well with people's normative visions of sustainability.

To avoid such limitations, ethical considerations should be explicitly woven throughout the process of index development rather than being considered after indexes are created, as in this project. The index development community has taken a step in this direction with the move to include stakeholders in the index development process. Participatory processes of index development are often construed to be a move toward including normative concerns in indexes as it is increasingly recognized that 1) local community values can shape priorities and subsequently what should be measured and 2) commitments to participatory justice theory entail local participation in decision-making. While these are some of the many reasons to include stakeholders in the index development process, merely including stakeholders as partners with technical experts on an index development team is not sufficient to ensure that normative concerns are fully considered in sustainability indexes. If professional index developers working with local communities are focusing on environmental and economic data as they often do (e.g. Fraser et al. 2006, Reed et al. 2008) then social aspects of sustainability including the aesthetic appreciation of ecosystems, the ability for cultural practices to continue, and justice within a community or nation may be overlooked. Furthermore, most index developers are simply not trained in ethics, philosophical analysis, or worldviews analysis and thus their guidelines for index development barely mention normative concerns. Involving ethicists in the process can help to overcome these potential limits.

Ethicists can provide a counterpoint to the focus on quantitative environmental and economic data favored by many index developers to ensure that cultural priorities for sustainability can also impact the index, whether through surveys of perceptions of environmental or social sustainability or through the use of new indicators devised to monitor cultural sustainability such as the percent of time rituals may be cancelled due to water or air pollution or other environmental degradation. Thus ethicists can not only help ensure such factors have a space in indexes but also can help identify ways to monitor social and ethical goals. Furthermore, at a broad level, they can help structure and make sense of the collaboration between academic experts, policy-makers, and stakeholders of a variety of worldviews through theories such as that of the broad categories or the work of Norton, Elling and others to which local stakeholders and traditional index developers will not necessarily have access, or even know about (Elling 2008, Norton 2005).

Ethicists using the theory of broad principles can facilitate the identification of such principles though rigorous processes of comparison among people of multiple worldviews. Fostering such collaboration is necessary since problems of climate change, desertification, species extinction, and environmental disease cross community and national boundaries and therefore require collaborative solutions involving people from multiple worldviews.

Finally, ethicists trained to identify normative priorities who are familiar with a variety of worldviews can help local communities articulate their norms, uncovering what is explicitly and implicitly valued. With such information coupled with knowledge of the interrelationships of social and environmental issues, sustainability ethicists can help identify when policies do not mesh with a groups' ideals. Such processes can help participants in index development identify indicators to include in or reject from indexes, threshold values beyond which an index is deemed unsustainable, weighting factors used to aggregate the indicators, or other technical details of indexes.

Since ethicists can aid the entire index development process I suggest a general framework for index development in which normative experts are involved throughout the process, working with technical experts, local stakeholders, and policy-makers (Bleicher and Gross 2010, Fraser et al. 2006, Holden 2011, Mickwitz and Melanen 2009, Rametsteiner et al. 2011, Reed et al. 2006). The precise relationship between these groups should be determined on a case-by-case basis to capitalize on existing resources and enable diverse participation in the process, but the general scope of such a process can be identified here.

First, locals, policy-makers, ethicists, and technical experts brainstorm priorities and indicators. Next, the potentially lengthy list of indicators must be reduced to a smaller, more manageable list, based on technical and normative considerations. Natural and social scientists would suggest indicators and aggregation methods while ethicists would work to ensure that all ethical priorities were included and suggest indicators and comment upon the aggregation methods. Locals and policy-makers could also be involved in this reduction process or they could wait to revise or approve the short list of indicators and indexes themselves as suggested by the index developers. If one prioritizes participatory justice then care must be taken to ensure that the priorities of locals are seriously considered. To do this, locals should be able to participate in the whole process if they so choose, ethicists should advocate for normative claims during index development, and index developers must be willing to revise indexes if requested by stakeholders for normative reasons. Once the indicators and index were agreed upon, they could be put into practice to monitor progress toward sustainability and the ramifications of policies over time.

Over time the process should be repeated as people recognize that the indexes or policies are not having the effect they desire, new situations arise, new knowledge is gained, or the broad principles are revised as new normative commitments are identified or come into the community with new members. Given this openness to revision, the indexes resulting from such a process should always be understood as the latest, strongest iteration of a sustainability index, but not a final or absolute truth.

Of course the method suggested here is not an easy solution. After all, it involves all of the challenges of interdisciplinary research about complex, dynamic systems; cooperative goal-setting among formally trained experts, lay experts, and policy-makers; and a process that in many ways runs counter to the status quo. But multiple signs indicate that the time is ripe for such a move. Indexes are increasingly used throughout society and their need for ethical assistance is becoming more visible. Some index theorists have recognized that normative priorities play a role in index development (Bleicher and Gross 2010: 603, Burger et al. 2010, Olalla-Tárraga 2006, Walter and Stuetzel 2009) and a few even recognize the need to consider diverse ethical perspectives to ensure that the indexes are ethically sound before they are adopted and more difficult to revise (Dahl 1997: 82). Additionally, the international community, as represented in *The Future We Want*, calls for an increased focus on indexes that are explicitly influenced by normative claims (United Nations: 2012a). Resources for such work are also increasing as index theory, sustainability studies, and environmental ethics have developed significantly in recent years and more students are trained in these interdisciplinary areas. Finally, lay people are increasingly involved in index formation and their own expertise regarding sustainability is increasingly recognized.

Efforts to recognize the role of normative priorities in such processes will enable people to understand what indexes do and do not monitor and thus determine whether or not an index needs revision to align with the priorities for the future found in the worldviews of those affected by it. With respect to sustainability of course, integrating these elements is necessary to ensure that sustainability indexes are able to monitor progress toward comprehensive visions of sustainability in which technical and normative considerations mutually influence each other and subsequently push society toward sustainability.

References

Agyeman, J. 2005. Alternatives for Community and Environment: Where Justice and Sustainability Meet. *Environment: Science and Policy for Sustainable Development*, 47, 10–23.
Agyeman, J. 2007. Communicating "Just Sustainability". *Environmental Communication*, 1, 119–122.
Agyeman, J., Bullard, R. D. & Evans, B. 2002. Exploring the Nexus: Bringing Together Sustainability, Environmental Justice and Equity. *Space & Polity*, 6, 77–90.
Agyeman, J., Bullard, R. D. & Evans, B. (eds) 2003. *Just Sustainabilities: Development in an Unequal World*, Cambridge, MA: The MIT Press.
Agyeman, J. & Warner, K. 2002. Putting "Just Sustainability" Into Place: From Paradigm to Practice. *Policy and Management Review*, 2, 8.
Ahlen, J. 2004. Examining the Relationship Between Energy and HDI. *Working Paper*, Boston University.
Aiken, B. 2001. The Earth Charter: Buddhist and Christian Approaches. *Buddhist-Christian Studies*, 21, 115–117.
Ammar, N. H. 1995. Islam: Population and the Environment: a Textual and Jursistic View. *In*: Coward, H. (ed.) *Population, Consumption, and the Environment: Religious and Secular Responses*. Albany: State University of New York Press.
Anderton, D. L., Anderson, A. B., Rossi, P. H., Oakes, J. M., Raser, M. R., Weber, E. W. & Calabrese, E. J. 1995. Studies Used to Prove Charges of Environmental Racism are Flawed. *In*: Petrikin, J. S. (ed.) *Environmental Justice: At Issue*. San Diego: Greenhaven Press.
Assefa, G. & Frostell, B. 2007. Social Sustainability and Social Acceptance in Technology Assessment: A Case Study of Energy Technologies. *Technology in Society*, 29, 63–78.
Astleithner, F. & Hamedinger, A. 2003. The Analysis of Sustainability Indicators as Socially Constructed Policy Instruments: Benefits and Challenges of "Interactive Research". *Local Environment*, 8, 627–640.
Attfield, R. & Wilkins, B. 1994. Sustainability. *Environmental Values*, 3, 155–158.
Awerbuch, S. & Sauter, R. 2006. Exploiting the Oil-GDP Effect to Support Renewables Deployment. *Energy Policy*, 34, 2805–2819.
Ayres, R. 1997. Forum: Comments on Georgescu-Roegen. *Ecological Economics*, 22, 285–287.
Ayres, R. U. & Nair, I. 1984. Thermodynamics and Economics. *Physics Today*, 62–71.
Azar, C., Holmberg, J. & Lindgren, K. 1996a. Methodological and Ideological Options: Socio-Ecological Indicators for Sustainability. *Ecological Economics*, 18, 89–112.

Azar, C., Holmberg, J. & Lindgren, K. 1996b. Socio-ecological Indicators for Sustainability. *Ecological Economics*, 18, 89–112.

Bahgat, G. 2006. Europe's Energy Security: Challenges and Opportunities. *International Affairs*, 82, 961–975.

Bakkes, J. 1997. Research Needs: Part One – Introduction. In: Moldan, B., Billharz, S. & Matravers, R. (eds) *Sustainability Indicators: a Report on the Project on Indicators of Sustainable Development*. New York: John Wiley & Sons.

Barbour, I. G. 2000. Scientific and Religious Perspectives on Sustainability. In: Hessel, D. T. & Ruether, R. R. (eds) *Christianity and Ecology: Seeking the Well-Being of Earth and Humans*. Cambridge, MA: Distributed by Harvard University Press for the Harvard University Center for the Study of World Religions.

Barbour, I., Brooks, H., Lakoff, S. & Opie, J. 1982. *Energy and the American Values*. New York: Praeger Publishers.

Barrett, C. B. & Grizzle, R. 1999. A Holistic Approach to Sustainability Based on Pluralism Stewardship. *Environmental Ethics*, 21, 23–42.

Bartelmus, P. 1997. Box 3G: Greening the National Accounts. In: Moldan, B., Billharz, S. & Matravers, R. (eds) *Sustainability Indicators: a Report on the Project on Indicators of Sustainable Development*. New York: John Wiley & Sons.

Bartelmus, P. 1999. *Greening the National Accounts: Approach and Policy Use*. New York: United Nations Department of Economic and Social Affairs, Discussion Paper No 3.

Bastianoni, S., Pulselli, F. M. & Tiezzi, E. 2004. The Problem of Assigning Responsibility for Greenhouse Gas Emissions. *Ecological Economics*, 49, 253–257.

Becker, J. 2004. Making Sustainable Development Evaluations Work. *Sustainable Development*, 12, 200–211.

Beckerman, W. 1994. Sustainable Development: is it a Useful Concept? *Environmental Values*, 3, 191–209.

Bell, S. 1999. *Sustainability Indicators: Measuring the Immesurable?* London: Earthscan.

Bell, S. & Morse, S. 2003. *Measuring Sustainability: Learning from Doing*. London: Earthscan.

Bennett, J. C. 1946. *Christian Ethics and Social Policy*, New York: Charles Scribner's Sons.

Berry, T. 1995. The Viable Human. In: Sessions, G. (ed.) *Deep Ecology for the Twenty-First Century*. Boston: Shambhala.

Binder, C. R., Feola, G. & Steinberger, J. K. 2010. Considering the Normative, Systemic and Procedural Dimensions in Indicator-based Sustainability Assessments in Agriculture. *Environmental Impact Assessment Review*, 30, 71–81.

Bleicher, A. & Gross, M. 2010. Sustainability Assessment and the Revitalization of Contaminated Sites: Operationalizing Sustainable Development for Local Problems. *International Journal of Sustainable Development and World Ecology*, 17, 57–66.

Blodgett, A. D. 2006. An Analysis of Pollution and Community Advocacy in "Cancer Alley": Setting an Example for the Environmental Justice Movement in St James Parish, Louisiana. *Local Environment*, 11, 647–661.

Börner, K. 2011. Science Maps as Visual Interfaces to Digital Libraries. Lecture given at the University of North Texas.

Bosivert, V., Holec, N. & Vivien, F.-D. 1998. Economic and Environmental Information for Sustainability. In: Faucheux, S. & O'Connor, M. (eds) *Valuation for Sustainable Development: Methods and Policy Indicators*. New York: John Wiley & Sons.

Bowen, W. 2002. An Analytical Review of Environmental Justice Research: What Do We Really Know? *Environmental Management*, 29, 3–15.

Brightman, E. S. 1933. *Moral Laws.* New York: Abingdon Press.
Brower, M. 1995. Comments on Stirling's "Diversity and Ignorance in Electricity Supply Investment". *Energy Policy,* 23, 115–116.
Brown, D. A. (ed.) 1994. *Proceedings on Interdisciplinary Conference Held at the United Nations on The Ethical Dimensions of the United Nations Program on Environment and Development Agenda 21.* New York: Earth Ethics Research Group Inc. Northeast Chapter.
Brown, D. A. & Lemons, J. (eds) 1995. *Sustainable Development: Science, Ethics, and Public Policy,* Dordrecht, The Netherlands: Kluwer Academic Publishers.
Bullard, R. D. 1990. *Dumping in Dixie: Race, Class, and Environmental Quality.* Boulder: Westview Press.
Bullard, R. D. 1993. *Confronting Environmental Racism: Voices from the Grassroots.* Boston, MA: South End Press.
Bullard, R. D. 1994a. Confronting Environmental Racism: Waste Trade and Agenda 21. In: Brown, D. A. (ed.) *Proceedings on Interdisciplinary Conference Held at the United Nations on The Ethical Dimensions of the United Nations Program on Environment and Development Agenda 21.* New York: Earth Ethics Research Group Inc. Northeast Chapter.
Bullard, R. D. 1994b. A New Chicken-or-Egg Debate: Which Comes First – The Neighborhood, or the Toxic Dump? *The Workbook,* 19, 60–62.
Bullard, R. D. 1994c. *Unequal Protection: Environmental Justice and Communities of Color.* San Francisco: Sierra Club Books.
Bullard, R. D. 1995. Decision Making. In: Westra, L. & Wenz, P. S. (eds) *Faces of Environmental Racism: Confronting Issues of Global Justice.* Lanham, MD: Rowman & Littlefield.
Bullard, R. D. 2000. *Dumping in Dixie: Race, Class, and Environmental Quality.* Boulder: Westview Press.
Burford, G., King, S., Knitter, P. & Mcdaniel, J. 1997. A Buddhist-Christian Contribution to the Earth Charter. *Buddhist-Christian Studies,* 17, 209–213.
Burger, P., Daub, C.-H. & Scherrer, Y. M. 2010. Creating Values for Sustainable Development. *International Journal of Sustainable Development & World Ecology,* 17, 1–3.
Callicott, J. B. 2010. Chapter 4. In: Raffaelle, R., Robison, W. & Selinger, E. (eds) *Sustainability Ethics: Five Questions.* Copenhagen: Automatic Press / VIP.
Caney, S. 2010a. Climate Change, Human Rights, and Moral Thresholds. In: Gardiner, S. M., Caney, S., Jamieson, D. & Shue, H. (eds) *Climate Ethics: Essential Readings.* Oxford: Oxford University Press.
Caney, S. 2010b. Cosmopolitan Justice, Responsibility, and Global Climate Change. In: Gardiner, S. M., Caney, S., Jamieson, D. & Shue, H. (eds) *Climate Ethics: Essential Readings.* Oxford: Oxford University Press.
Capek, S. M. 1993. The "Environmental Justice" Frame: A Conceptual Discussion and an Application. *Social Problems,* 40, 5–24.
Carlisle, E. 1972. The Conceptual Structure of Social Indicators. In: Schonfield, A. & Shaw, S. (eds) *Social Indicators and Social Policy.* London: Heinemann Educational Books.
Carson, R. 2002. *Silent Spring.* New York: Houghton Mifflin Company.
Chapman, A. R. 2000. Science, Religion, and the Environment. In: Chapman, A. R., Peterson, R. L. & Smith-Moran, B. (eds) *Consumption, Population, and Sustainability.* Washington, D.C.: Island Press.
Chavez, C. 1993. Farm Workers at Risk. In: Hofrichter, R. (ed.) *Toxic Struggles: The Theory and Practice of Environmental Justice.* Philadelphia: New Society Publishers.

Checker, M. A. 2002. "It's in the Air": Redefining the Environment as a New Metaphor for Old Social Justice Struggles. *Human Organization*, 61, 94.

Checker, M. 2007. "But I Know It's True": Environmental Risk Assessment, Justice, and Anthropology. *Human Organization*, 66, 112.

Chishti, S. K. K. 2003. Fitra: An Islamic Model for Humans and the Environment. In: Foltz, R. C., Denny, F. M. & Bahruddin, A. (eds) *Islam and Ecology: A Bestowed Trust*. Cambridge, MA: Harvard University Press.

Clark, C. W. 1997. Forum: Renewable Resources and Economic Growth. *Ecological Economics*, 22, 275–276.

Clarke, L. 2003. The Universe Alive: Nature in the Masnavi of Jalal al-Din Rumi. In: Foltz, R. C., Denny, F. M. & Bahruddin, A. (eds) *Islam and Ecology: A Bestowed Trust*. Cambridge, MA: Harvard University Press.

Cobb, C. W. & Halstead, T. 1994. *The Genuine Progress Indicator*. San Francisco: Redefining Progress.

Cole, L. W. & Foster, S. R. 2001. *From the Ground Up: Environmental Racism and the Rise of the Environmental Movement*. New York: New York University Press.

Commission for Racial Justice United Church of Christ. 1987. Toxic Wastes and Race in the United States: A National Report on the Racial and Socio-Economic Characteristics of Communities with Hazardous Waste Sites.

Common, M. 1997. Forum: Is Georgescu-Roegen Versus Solow/Stiglitz the Important Point? *Ecological Economics*, 22, 277–279.

Conrad, S. M. 2011. A Restorative Environmental Justice for Prison E-Waste Recycling. *Peace Review*, 23, 348–355.

Corburn, J. 2005. *Street Science: Community Knowledge and Environmental Health Justice*. Cambridge, MA: MIT Press.

Costantini, V., Gracceva, F., Markandya, A. & Vicini, G. 2007. Security of Energy Supply: Comparing Scenarios from a European Perspective. *Energy Policy*, 35, 210–226.

Coulson, N. J. 2003. *A History of Islamic Law*. Edinburgh: Edinburgh University Press.

Council, N. R. 2000. *Ecological Indicators for the Nation*. Washington, D.C.: National Academy Press.

Court, T. D. L. 1990. *Beyond Brundtland: Green Development in the 1990s*. Atlantic Highlands, NJ: Zed Books.

Custance, J. 2002. The Development of National, Regional, and Local Indicators of Sustainable Development in the United Kingdom. *Statistical Journal of the United Nations ECE*, 19, 19–28.

Dahl, A. L. 1997. The Big Picture: Comprehensive Approaches Part One – Introduction. In: Moldan, B., Billharz, S. & Matravers, R. (eds) *Sustainability Indicators: a Report on the Project on Indicators of Sustainable Development*. New York: John Wiley & Sons.

Daly, H. E. 1997a. Forum: Georgescu-Roegen versus Solow/Stiglitz. *Ecological Economics*, 22, 261–266.

Daly, H. E. 1997b. Forum: Reply to Solow/Stiglitz. *Ecological Economics*, 22, 271–273.

Daly, H. E., Cobb, J. B. & Cobb, C. W. 1989. *For the Common Good: Redirecting the Economy toward Community, the Environment, and a Sustainable Future*. Boston: Beacon Press.

Davidsdottir, B., Basoli, D., Fredericks, S. E. & Enterline, C. L. 2007. Measuring Sustainable Energy Development: the Development of a Three-Dimensional Index. In: Erickson, J. D. & Gowdy, J. (eds) *Frontiers in Environmental Valuation and Policy*. Cheltenham, UK: Edward Elgar.

Deats, P. 1986. Conflict and Reconciliation in Communitarian Social Ethics. In: Deats, P. & Robb, C. (eds) *The Boston Personalist Tradition*. Macon, GA: Mercer.

Dell, R. M. & Rand, D. A. J. 2001. Energy Storage – a Key Technology for Global Energy Sustainability. *Journal of Power Sources*, 100, 2–17.

Derr, T. S. 2000. Global Eco-Logic. *First Things*, February, 9–12.

DOE/EIA. 2006. *Emissions of Greenhouse Gases in the United States 2005, DOE/EIA-0573 (2005)*. Available: ftp://ftp.eia.doe.gov/pub/oiaf/1605/cdrom/pdf/ggrpt/057305.pdf [Accessed March 18, 2013].

Dorsey, J. W. 2009. Restorative Environmental Justice: Assessing Brownfield Initiatives, Revitalization, and Community Economic Development in St. Petersburg, Florida. *Environmental Justice*, 2, 69–80.

Du Pisani, J. A. 2006. Sustainable Development – Historical Roots of the Concept. *Environmental Sciences*, 3, 86–96.

Earth Charter Initiative. A Short History of the Earth Charter. Available: http://www.earthcharterinaction.org/download/about_the_Initiative_history_2t.pdf [Accessed March 7, 2013].

Ehrlich, P. & Ehrlich, A. 1968. *The Population Bomb*. New York: Ballantine Books.

Elling, B. 2008. *Rationality and the Environment*. London: Earthscan.

Emerson, J. W., Hsu, A., Levy, M. A., De Sherbinin, A., Mara, V., Esty, D. C. & Jaiteh, M. 2012a. *2012 Environmental Performance Index and Pilot Trend Environmental Performance Index*. New Haven, CT: Yale Center for Environmental Law and Policy.

Emerson, J. W., Hsu, A., Levy, M. A., De Sherbinin, A., Mara, V., Esty, D. C. & Jaiteh, M. 2012b. *Environmental Performance Index: Future Work*. New Haven, CT: Yale University. Available: http://epi.yale.edu/epi2012/futurework [Accessed March 7, 2013].

Emmett, E. A. & Desai, C. 2010. Community First Communication: Reversing Information Disparities to Achieve Environmental Justice. *Environmental Justice*, 3, 79–84.

Energy Information Administration. 2006. *International Energy Outlook 2006, DOE/EIA-0484 (2006)*. Washington, DC: EIA.

Energy Information Administration. 2009. *Emissions of Greenhouse Gases Report*. Washington, DC: EIA.

Engel, J. R. & Engel, J. G. 1990. *Ethics of Environment and Development: Global Challenge, International Response*. Tucson: University of Arizona Press.

England, R. W. 2001. Alternatives to Gross Domestic Product: A Critical Survey. In: Cleveland, C. J., Stern, D. I. & Costanza, R. (eds) *The Economics of Nature and the Nature of Economics*. Northampton, MA: Edward Elgar.

Environmental Protection Agency. EJView. Available: http://epamap14.epa.gov/ejmap/entry.html [Accessed March 7, 2013].

European Commission. 2012. *Sustainable Development*. Available: http://ec.europa.eu/environment/eussd/ [Accessed March 7, 2013].

Eurostat. 2012a. *About Eurostat: Introduction*. European Commission. Available: http://epp.eurostat.ec.europa.eu/portal/page/portal/about_eurostat/introduction [Accessed March 7, 2013].

Eurostat. 2012b. *Statistics*. European Commission. Available: http://epp.eurostat.ec.europa.eu/portal/page/portal/statistics/themes [Accessed March 7, 2013].

Eurostat. 2012c. *Sustainable Development Indicators: Context*. European Commission. Available: http://epp.eurostat.ec.europa.eu/portal/page/portal/sdi/context [Accessed March 7, 2013].

Eurostat. 2012d. *Sustainable Development Indicators: Indicators*. European Commission. Available: http://epp.eurostat.ec.europa.eu/portal/page/portal/sdi/indicators [Accessed March 7, 2013].

Evans, G. W. & Marcynyszyn, L. A. 2004. Environmental Justice, Cumulative Environmental Risk, and Health Among Low- and Middle-Income Children in Upstate New York. *American Journal of Public Health*, 94, 1942–1944.

Failing, L. & Gregory, R. 2003. Ten Common Mistakes in Designing Biodiversity Indicators for Forest Policy. *Journal of Environmental Management*, 68, 121–132.

FAO Council. 1988. *Aspects of FAO's Policies, Programmes, Budget and Activities Aimed at Contributing to Sustainable Development*. Rome: FAO.

Faruqui, N. I. 2001. Islam and Water Management: Overview and Principles. In: Faruqui, N. I., Biswas, A. K. & Bino, M. J. (eds) *Water Management in Islam*. New York: United Nations University Press.

Faruqui, N. I., Biswas, A. K. & Bino, M. J. (eds) 2001. *Water Management in Islam*. New York: United Nations University Press.

Faucheux, S. & O'Connor, M. (eds) 1998. *Valuation for Sustainable Development: Methods and Policy Indicators*. New York: John Wiley & Sons.

Feibleman, J. K. 1969. *An Introduction to the Philosophy of Charles S. Peirce: Interpreted as a System*. Cambridge, MA: MIT Press.

Fernandez-Sanchez, G. & Rodriguez-Lopez, F. 2010. A Methodology to Identify Sustainability Indicators in Construction Project Management – Application to Infrastructure Projects in Spain. *Ecological Indicators*, 10, 1193–1201.

Figueroa, R. 2002. Other Faces: Latinos and Environmental Justice. In: Westra, L. & Lawson, B. (eds) *Faces of Environmental Racism: Confronting Issues of Global Justice*. 2nd ed. Lanham, MD: Rowman & Littlefield.

Figueroa, R. & Mills, C. 2001. Environmental Justice. In: Jamieson, D. (ed.) *A Companion to Environmental Philosophy*. Malden, MA: Basil Blackwell.

Figueroa, R. M. & Waitt, G. 2010. Climb: Restorative Justice, Environmental Heritage, and the Moral Terrains of Uluru-Kata Tjuta National Park. *Environmental Philosophy*, 7, 135–163.

Foltz, R. C. 2003. Islamic Environmentalism: A Matter of Interpretation. In: Foltz, R. C., Denny, F. M. & Bahruddin, A. (eds) *Islam and Ecology: A Bestowed Trust*. Cambridge, MA: Harvard University Press.

Ford, L. 2012. *Rio+20 Politicians Deliver "New Definition of Hypocrisy" Claim NGOs*. Available: http://www.guardian.co.uk/global-development/2012/jun/21/rio20-politicians-hypocrisy-ngos [Accessed March 7, 2013].

Forstner, H. 1997. Box 3J: Sustainable Industry Indicators: The DSR Framework and Beyond. In: Moldan, B., Billharz, S. & Matravers, R. (eds) *Sustainability Indicators: a Report on the Project on Indicators of Sustainable Development*. New York: John Wiley & Sons.

Fraser, E. D. G., Dougill, A. J., Mabee, W. E., Reed, M. & Mcalpine, P. 2006. Bottom Up and Top Down: Analysis of Participatory Processes for Sustainability Indicator Identification as a Pathway to Community Empowerment and Sustainable Environmental Management. *Journal of Environmental Management*, 78, 114–127.

Fredericks, S. E. 2011. Monitoring Environmental Justice. *Environmental Justice*, 4(1), 63–69.

Fredericks, S. E. 2012a. Agenda 21. In: Jenkins, W. & Bauman, W. (eds) *Berkshire Encyclopedia of Sustainability* (Vol. 6: Measurements, Indicators, and Research Methods for Sustainability). Great Barrington, MA: Berkshire Publishing Co.

Fredericks, S. E. 2012b. Challenges to Measuring Sustainability. *In:* Spellerberg, I., Fogel, D., Fredericks, S. E., Butler Harrington, L. M., Pronto, M. & Wouters, P. (eds) *Berkshire Encyclopedia of Sustainability*. Great Barrington, MA: Berkshire Publishing Co.

Fredericks, S. E. 2012c. Justice in Sustainability Indexes and Indicators. *International Journal of Sustainable Development & World Ecology*, 19(6), 490–499.

Gallego Carrera, D. & Mack, A. 2010. Sustainability Assessment of Energy Technologies via Social Indicators: Results of a Survey among European Energy Experts. *Energy Policy*, 38, 1030–1039.

Gallopin, G. C. 1997. Indicators and their Use: Information for Decision-making Part One – Introduction. *In:* Moldan, B., Billharz, S. & Matravers, R. (eds) *Sustainability Indicators: a Report on the Project on Indicators of Sustainable Development*. New York: John Wiley & Sons.

Geczi, E. 2007. Sustainability and Public Participation: Toward an Inclusive Model of Democracy. *Administrative Theory & Praxis*, 29, 375–393.

Gibbs, L. M. 1999. *Love Canal: The Story Continues ...* Boston: South End Press.

Gibson, R. B. 2005. *Sustainability Assessment: Criteria and Processes*. London: Earthscan.

Gillroy, J. M. & Bowersox, J. (eds) 2002. *The Moral Austerity of Decision Making: Sustainability, Democracy, and Normative Argument in Policy and Law*. Durham: Duke University Press.

Gingerich, E. F. R. 2010. Global Reporting Initiative. *In:* Laszlo, C., Christensen, K., Fogel, D., Wagner, T. & Whitehouse, P. (eds) *The Berkshire Encyclopedia of Sustainability*. Great Barrington, MA: Berkshire Publishing Co.

Goldman, B. A. & Fitton, L. 1994. *Toxic Wastes and Race Revisited: An Update of the 1987 Report on the Racial and Socioeconomic Characteristics of Communities with Hazardous Waste Sites*. Washington, DC: Center for Policy Alternatives, National Association for the Advancement of Colored People, United Church of Christ Commission for Racial Justice.

Goldstein, B. D. 2005. Advances in Risk Assessment and Communication. *Annual Review of Public Health*, 26, 141–163.

Gorovitz Robertson, H. 2008. Controlling Existing Facilities. *In:* Gerrard, M. & Foster, S. R. (eds) *The Law of Environmental Justice: Theories and Procedures to Address Disproportionate Risks*. Chicago: American Bar Association, Section of Environment, Energy and Resources.

Gottleib, R. 1993. *Forcing the Spring: The Transformation of the American Environmental Movement*. Washington, D.C.: Island Press.

Gray, R. H. 1994. Corporate Reporting for Sustainable Development: Accounting for Sustainability in 2000 AD. *Environmental Values*, 3, 17–45.

Grubb, M., Butler, L. & Twomey, P. 2006. Diversity and Security in UK Electricity Generation: The Influence of Low-carbon Objectives. *Energy Policy*, 34, 4050–4062.

Hák, T., Moldan, B. & Dahl, A. L. 2007. *Sustainability Indicators: A Scientific Assessment*. Washington, D.C.: Island PRess.

Hallaq, W. B. 1995a. *Law and Legal Theory in Classical and Medieveal Islam*. Brookfield, VT: Ashgate Publishing.

Hallaq, W. B. 1995b. Was the Gate of Ijtihād Closed? *In: Law and Legal Theory in Classical and Medieval Islam*. Brookfield, VT: Ashgate Publishing.

Hamed, S.-E. A. 2003. Capacity Building for Sustainable Development: The Dilemma of Islamization of Environmental Institutions. *In:* Foltz, R. C., Denny, F. M. &

Bahruddin, A. (eds) *Islam and Ecology: A Bestowed Trust*. Cambridge, MA: Harvard University Press.

Haq, S. N. 2003. Islam and Ecology: Toward Retrieval and Reconstruction. In: Foltz, R. C., Denny, F. M. & Bahruddin, A. (eds) *Islam and Ecology: A Bestowed Trust*. Cambridge, MA: Harvard University Press.

Harding, A. (ed.) 2007. *Access to Environmental Justice: A Compartive Study*. Boston: Martinus Nijhoff Publishers.

Hargrove, E. C. 1985. The Role of Rules in Ethical Decision Making. *Inquiry*, 28, 3–42.

Hargrove, E. C. 1997. Environmental Ethics and the Earth Charter. *Environmental Ethics*, 19(1), 3–4.

Harlan, S. L., Brazel, A. J., Jenerette, G. D., Jones, N. S., Larsen, L., Prashad, L. & Stefanov, W. L. 2008. In the Shade of Affluence: The Inequitable Distribution of the Urban Heat Island. In: Wilkinson, R. C. & Fruedenburg, W. R. (eds) *Equity and the Environment*. Amsterdam: Elsevier.

Hartshorne, C. & Weiss, P. (eds) 1931. *Collected Papers of Charles Sanders Peirce*. Cambridge, MA: Harvard University Press.

Harvey, F. June 25, 2012. Rio +20 Makes no Fresh, Green Breast of the New World. *Rio+20 Earth Summit*. Available: http://www.guardian.co.uk/environment/blog/2012/jun/25/rio-20-great-gatsby [Accessed March 7, 2013].

Hawley, C. 2012. *Rio+20 has Become the Summit of Futility*. Spiegel Online International. Available: http://www.spiegel.de/international/world/widespread-criticism-of-rio-environment-summit-a-840181.html [Accessed March 7, 2013].

Hecq, W. J. 1997. Box 2I: From Static to Dynamic Indicators of Sustainable Development: Example of the Economy-Energy-Environment Link. In: Moldan, b., Billharz, S. & Matravers, R. (eds) *Sustainability Indicators: a Report on the Project on Indicators of Sustainable Development*. New York: John Wiley & Sons.

Hepple, R. P. & Benson, S. M. 2005. Geological Storage of Carbon Dioxide as a Climate Change Mitigation Strategy: Performance Requirements and the Implications of Surface Seepage. *Environmental Geology*, 47, 576–585.

Heredia, R. C. 1994. The Ethical Implicationsn of a Global Climate Change – A Third World Perspective. In: Brown, D. (ed.) *Proceedings on Interdisciplinary Conference Held at the United Nations on the Ethical Dimensions of the United Nations Program on Environment and Development, Agenda 21*. New York: Earth Ethics Research Group, Northeast Chapter, United Nations.

Herzog, H. J. & Drake, E. M. 1996. Carbon Dioxide Recovery and Disposal from Large Energy Systems. *Annual Review of Energy and the Environment*, 21, 145–166.

Hess, D. & Winner, L. 2007. Enhancing Justice and Sustainability at the Local Level: Affordable Policies for Urban Governments. *Local Environment*, 12, 379–395.

Heyd, T. 1994. Agenda 21 and the Limits of Technical Rationality. In: Brown, D. (ed.) *Proceedings on Interdisciplinary Conference Held at the United Nations on the Ethical Dimensions of the United Nations Program on Environment and Development, Agenda 21*. New York: Earth Ethics Research Group, Northeast Chapter, United Nations.

Hezri, A. A. & Dovers, S. R. 2006. Sustainability Indicators, Policy and Governance: Issues for Ecological Economics. *Ecological Economics*, 60, 86–99.

Hezri, A. A. & Hasan, M. N. 2004. Management Framework for Sustainable Development Indicators in the State of Selangor, Malaysia. *Ecological Indicators*, 4, 287–304.

Hitchcock, D. & Willard, M. 2009. *The Business Guide to Sustainability: Practical Strategies and Tools for Organizations*. London: Earthscan.

Holden, M. 2011. Public Participation and Local Sustainability: Questioning a Common Agenda in Urban Governance. *International Journal of Urban and Regional Research*, 35, 312–329.

Holloway, S. 2001. Storage of Fossil Fuel-Derived Carbon Dioxide Beneath the Surface of the Earth. *Annual Review of Energy and the Environment*, 26, 145–166.

Holub, H. W., Tappeiner, G. & Tappenier, U. 1999. Commentary: Some Remarks on the "System of Integrated Environmental and Economic Accounting" of the United Nations. *Ecological Economics*, 29, 329–336.

Houghton, J. 2004. *Global Warming: the Complete Briefing*. New York: Cambridge University Press.

Hueting, R. 1991. Correcting National Income for Environmental Losses: a Practical Solution for a Theoretical Dilemma. In: Costanza, R. (ed.) *Ecological Economics: the Science and Management of Sustainability*. New York: Columbia University Press.

Hueting, R., Bosch, P. R. & De Boer, B. 1992. *Methodology for the Calculation of Sustainable National Income*. Gland, Switzerland: World Wide Fund for Nature.

Hueting, R. & Reijnders, L. 2004. Broad Sustainability Contra Sustainability: The Proper Construction of Sustainabiltiy Indicators. *Ecological Economics*, 50.

Hunold, C. & Young, I. 1998. Justice, Democracy, and Hazardous Siting. *Political Studies*, XLVI, 82–95.

IAEA/IEA. 2001. *Indicators for Sustainable Energy Development*. 9th Session of the United Nations Commission on Sustainable Development, April.

IEA/OECD. 2001. *Towards a Sustainable Energy Future*. Paris, France: OECD.

Inhaber, H. 1976. *Environmental Indices*. New York: John Wiley & Sons.

Intergovernmental Panel on Climate Change. 2007. Climate Change 2007: The Physical Science Basis: Summary Report for Policy Makers. Paris: Intergovernmental Panel on Climate Change.

International Institute for Sustainable Development. 2012. *Compendium of Sustainable Development Indicator Initiatives*. Available: http://www.iisd.org/measure/compendium/ [Accessed March 7, 2013].

Johnston, L. 2010. The Religious Dimensions of Sustainability: Institutional Religions, Civil Society, and International Politics since the Turn of the Twentieth Century. *Religion Compass*, 4, 176–189.

Johnston, L. 2013. *Religion and Sustainability: Social Movements and the Politics of the Environment*. Bristol, CT: Equinox Publishing Ltd.

Johnston, L. F. & Snyder, S. 2011. Practically Natural: Religious Resources for Environmental Pragmatism. In: Bauman, W. A., Bonhannon II, R. R. & O'Brien, K. J. (eds) *Inherited Land: the Changing Grounds of Religion and Ecology*. Eugene, OR: Pickwick Publications.

Kamieniecki, S., Coleman, S. D. & Vos, R. O. 1995. The Effectiveness of Radical Environmentalists. In: Taylor, B. R. (ed.) *Ecological Resistance Movements: The Global Emergence of Radical and Popular Environmentalism*. Albany: State University of New York Press.

Katz, E. 1994. Sustainable Development and Imperialism: Ethical Reflections on Agenda 21. In: Brown, D. (ed.) *Proceedings on Interdisciplinary Conference Held at the United Nations on the Ethical Dimensions of the United Nations Program on Environment and Development, Agenda 21*. New York: Earth Ethics Research Group, Northeast Chapter, United Nations.

Kettani, A. 1984. Science and Technology in Islam: the Underlying Value System. In: Sardar, Z. (ed.) *The Touch of Midas: Science, Values and Environment in Islam and the West*. Manchester, NH: Manchester University Press.

Krank, S. & Wallbaum, H. 2011. Lessons from Seven Sustainability Indicator Programs in Developing Countries of Asia. *Ecological Indicators*, 11, 1385–1395.

Kraus, C. 1993. Blue-Collar Women and Toxic Waste Protests. In: Hofrichter, R. (ed.) *Toxic Struggles: The Theory and Practice of Environmental Justice*. Philadelphia: New Society Publishers.

Krieg, E. J. 1998. The Two Faces of Toxic Waste: Trends in the Spread of Environmental Hazards. *Sociological Forum*, 13, 3–20.

Landrigan, P. J., Raugh, V. A. & Galvez, M. P. 2010. Environmental Justice and the Health of Children. *Mount Sinai Journal of Medicine*, 77, 178–187.

Lefever, E. W. 1979. *Amsterdam to Nairobi: The World Council of Churches and the Third World*. Washington, D.C.: Ethics and Public Policy Center of Georgetown University.

Leveness, F. P. & Primeaux, P. D. 2004. Vicarious Ethics: Politics, Business, and Sustainable Development. *Journal of Business Ethics*, 51, 185–198.

Levett, R. 1998. Sustainability Indicators – Integrating Quality of Life and Environmental Protection. *Journal of the Royal Statistical Society Series A (Statistics in Society)*, 161, 291–302.

Litt, J. S., Tran, N. L. & Burke, T. A. 2002. Examining Urban Brownfields through the Public Health "Macroscope". *Environmental Health Perspectives*, 110, 183–193.

Llewellyn, O. A.-A.-R. 1984. Islamic Jurisprudence and Environmental Planning. *Journal of Research in Islamic Economics*, 1, 27–46.

Llewellyn, O. A.-A.-R. 1992. Desert Reclamation and Conservation in Islamic Law. In: Khalid, F. M. & O'Brien, J. (eds) *Islam and Ecology*. New York: Cassell Publishers.

Llewellyn, O. A.-A.-R. 2003. The Basis for a Discipline of Islamic Environmental Law. In: Foltz, R. C., Denny, F. M. & Bahruddin, A. (eds) *Islam and Ecology: A Bestowed Trust*. Cambridge, MA: Harvard University Press.

Lopez-Ridaura, S., Masera, O. & Astier, M. 2002. Evaluating the Sustainability of Complex Socio-Environmental Systems. The MESMIS framework. *Ecological Indicators*, 2, 135–148.

Low, N. & Gleeson, B. 1998. Situating Justice in the Environment: The Case of the BHP at the OK Tedi Copper Mine. *Antipode*, 30, 201–226.

Lucas, N., Price, T. & Tompkins, R. 1995. Diversity and Ignorance in Electricity Supply Investment: A Reply to Andrew Stirling. *Energy Policy*, 23, 5–16.

Maantay, J. 2002. Mapping Environmental Injustices: Pitfalls and Potential of Geographic Information Systems in Assessing Environmental Health and Equity. *Environmental Health Perspectives*, 110, 161–171.

Majeed, A. B. A. 2003. Islam in Malaysia's Planning and Development Doctrine. In: Foltz, R. C., Denny, F. M. & Bahruddin, A. (eds) *Islam and Ecology: A Bestowed Trust*. Cambridge, MA: Harvard University Press.

Marc, D. S. 2005. Equity and Information: Information Regulation, Environmental Justice, and Risks from Toxic Chemicals. *Journal of Policy Analysis and Management*, 24, 373.

Margolis, M. 2012. *The Rio+20 Conference Went From Good Intentions to the To-Do List From Hell*. The Daily Beast. Available: http://www.thedailybeast.com/articles/2012/06/23/the-rio-20-conference-went-from-good-intentions-to-the-to-do-list-from-hell.html [Accessed March 7, 2013].

Martin, M. C. 2002. Expanding the Boundaries of Environmental Justice: Native Americans and the South Lawrence Trafficway. *Policy and Management Review*, 2, 62.

Martin-Schramm, J. B. & Stivers, R. L. 2003. *Christian Environmental Ethics: A Case Method Approach*. Maryknoll, NY: Orbis Books.

Mascarenhas, A., Coelho, P., Subtil, E. & Ramos, T. B. 2010. The Role of Common Local Indicators in Regional Sustainability Assessment. *Ecological Indicators*, 10, 646–656.

McMahon, S. K. 2002. The Development of Quality of Life Indicators: a Case Study from the City of Bristol, UK. *Ecological Indicators*, 2, 177–185.

Meadows, D. H., Meadows, D. L., Randers, J. & Behrens III, W. W. 1972. *The Limits to Growth: a Report for the Club of Rome's Project on the Predicament of Mankind*. New York: Universe Books.

Mennis, J. 2002. Using Geographic Information Systems to Create and Analyze Statistical Surfaces of Population and Risk for Environmental Justice Analysis. *Social Science Quarterly (Blackwell Publishing Limited)*, 83, 281–297.

Mickwitz, P. & Melanen, M. 2009. The Role of Co-operation between Academia and Policymakers for the Development and use of Sustainability Indicators – a Case from the Finnish Kymenlaakso Region. *Journal of Cleaner Production*, 17, 1086–1100.

Mitcham, C. 1995. The Concept of Sustainable Development: Its Origins and Ambivalence. *Technology in Society*, 17, 311–326.

Mitra, A. 2003. A Tool for Measuring Progress: the Growing Popularity of Sustainable Indicators in the United States. *National Civic Review*, Fall, 30–45.

Moffatt, I. 1992. The Evolution of the Sustainable Development Concept: A Perspective from Australia. *Australian Geographical Studies*, 30, 27–42.

Mohai, P. & Bryant, B. 1995. Demographic Studies Reveal a Pattern of Environmental Injustice. In: Petrikin, J. S. (ed.) *Environmental Justice: At Issue*. San Diego: Greenhaven Press.

Moldan, B. 1997. Box 2D: Values, Services and Goods of the Geobiosphere. In: Moldan, B., Billharz, S. & Matravers, R. (eds) *Sustainability Indicators: a Report on the Project on Indicators of Sustainable Development*. New York: John Wiley & Sons.

Moldan, B., Billharz, S. & Matravers, R. (eds) 1997. *Sustainability Indicators: a Report on the Project on Indicators of Sustainable Development*. New York: John Wiley & Sons.

Mortensen, L. F. 1997. The Driving Force–State–Response Framework used by CSD. In: Moldan, B., Billharz, S. & Matravers, R. (eds) *Sustainability Indicators: a Report on the Project on Indicators of Sustainable Development*. New York: John Wiley & Sons.

Moses, M. 1993. Farmworkers and Pesticides. In: Bullard, R. D. (ed.) *Confronting Environmental Racism: Voices from the Grassroots*. Boston: South End Press.

Muelder, W. G. 1959. *Foundations of the Responsible Society*. New York: Abingdon Press.

Muelder, W. G. 1966. *Moral Law in Christian Social Ethics*. Richmond: John Knox Press.

Muelder, W. G. 1983. Christian Bases of Morality and Ethics for Today (Choosing Responsibly in a Revolutionary World). In: *The Ethical Edge of Christian Theology: Forty Years of Communitarian Personalism*. New York: E. Mellen Press.

Naess, A. 1973. The Shallow and The Deep, Long-Range Ecology Movement: a Summary. *Inquiry*, 16, 95–100.

Naess, A. 1989. *Ecology, Community and Lifestyle: Outline of an Ecosophy*. New York: Cambridge University Press.

Naess, A. 1995a. The Deep Ecological Movement: Some Philosophical Aspects. In: Sessions, G. (ed.) *Deep Ecology for the Twenty-First Century*. Boston: Shambhala.

Naess, A. 1995b. The Deep Ecology "Eight Points" Revisited. In: Sessions, G. (ed.) *Deep Ecology for the Twenty-First Century*. Boston: Shambhala.

Naess, A. 1995c. Self-Realization: An Ecological Approach to Being in the World. In: Sessions, G. (ed.) *Deep Ecology for the Twenty-First Century*. Boston: Shambhala.

Naess, A. 1995d. The Shallow and the Deep, Long-Range Ecology Movements: A Summary. In: Sessions, G. (ed.) *Deep Ecology for the Twenty-First Century*. Boston: Shambhala.

Najam, A. & Cleveland, C. 2003. Energy and Sustainable Development at Global Environmental Summits: An Evolving Agenda. *International Journal of Environment and Sustainability*, 5, 117–138.

Narain, S. July 4, 2012. Rio+20: Why It Failed? *The Blog*. Available from: http://www.huffingtonpost.com/sunita-narain/rio20-why-it-failed_b_1648399.html [Accessed March 7, 2013].

Nash, J. A. 1989. Ecological Integrity and Christian Political Responsibility. *Theology and Public Policy*, 1, 32–48.

Nash, J. A. 1991. *Loving Nature: Ecological Integrity and Christian Responsibility*. Nashville: Abingdon Press.

Nash, J. A. 1992. Ethical Concerns for the Global-Warming Debate. *Christian Century*, 109, 773–776.

Nash, J. A. 1993. Biotic Rights and Human Ecological Responsibilities. *Annual of the Society of Christian Ethics*, 137–162.

Nash, J. A. 1994. Ethics and the Economics-Ecology Dilemma: Toward a Just, Sustainable, and Frugal Future. *Theology and Public Policy*, 6, 33–63.

Nash, J. A. 1995. Toward the Revival and Reform of the Subversive Virtue: Frugality. *Annual of the Society of Christian Ethics*, 137–160.

Nash, J. A. 1996. The Politician's Moral Dilemma: The Moral Possibilities and Limits of Political Leadership in Confronting the Ecological Crisis. *CTNS Bulletin*, 16, 7–15.

Nash, J. A. 2000. Seeking Moral Norms in Nature: Natural Law and Ecological Responsibility. In: Hessel, D. T. & Ruether, R. R. (eds) *Christianity and Ecology: Seeking the Well-Being of Earth and Humans*. Cambridge, MA: Distributed by Harvard University Press for the Harvard University Center for the Study of World Religions.

Nash, J. A. 2001. Healing an Ailing Alliance: Ethics and Science face the Ambiguities of Water. *Journal of Faith and Science Exchange*, 5, 111–124.

Nasr, S. H. 1993. *The Need for a Sacred Science*. Albany: State University of New York Press.

Nasr, S. H. 1996. *Religion and the Order of Nature*. New York: Oxford University Press.

Nasr, S. H. 2000. Islam, the Contemporary Islamic World, and the Environmental Crisis. In: Hessel, D. T. & Ruether, R. R. (eds) *Christianity and Ecology: Seeking the Well-Being of Earth and Humans*. Cambridge, MA: Distributed by Harvard University Press for the Harvard University Center for the Study of World Religions.

National Research Council (US) Committee on Environmental Epidemiology 1991. *Environmental Epidemiology: Public Health and Hazardous Wastes*, Washington, D.C.: National Academy Press.

Nations, U. 2012. What is Rio+20?. Available: http://www.uncsd2012.org/ [Accessed March 7, 2013].

Nattrass, B. & Altomare, M. 2001. *The Natural Step for Business: Wealth, Ecology and the Evolutionary Corporation.* Gabriola Island, BC: New Society Publishers.

Neville, R. C. 1987. *The Puritan Smile: A Look Toward Moral Reflection.* Albany: State University of New York Press.

Neville, R. C. 1989. *Recovery of the Measure: Interpretation and Nature.* Albany: State University of New York Press.

Neville, R. C. 1992. *The Highroad Around Modernism.* Albany: State University of New York Press.

Neville, R. C. 1995. *Normative Cultures.* Albany: State University of New York Press.

Neville, R. C. (ed.) 2001a. *The Human Condition.* Albany: State University of New York Press.

Neville, R. C. (ed.) 2001b. *Religious Truth.* Albany: State University of New York Press.

Neville, R. C. (ed.) 2001c. *Ultimate Realities.* Albany: State University of New York Press.

Neville, R. C. & Wildman, W. J. 2001a. Introduction. In: Neville, R. C. (ed.) *Religious Truth: A Volume in the Comparative Religious Ideas Project.* Albany: State University of New York Press.

Neville, R. C. & Wildman, W. J. 2001b. Introduction. In: Neville, R. C. (ed.) *Ultimate Realities: A Volume in the Comparative Religious Ideas Project.* Albany: State University of New York Press.

Neville, R. C. & Wildman, W. J. 2001c. On Comparing Religious Ideas. In: Neville, R. C. (ed.) *Ultimate Realities.* Albany: State University of New York Press.

Norgaard, K. M. 2007. The Politics of Invasive Weed Management: Gender, Race, and Risk Perception in Rural California. *Rural Sociology,* 72, 450.

Norton, B. 1991. *Toward Unity among Environmentalists.* New York: Oxford University Press.

Norton, B. 1992. Sustainability, Human Welfare and Ecosystem Health. *Environmental Values,* 1, 97–111.

Norton, B. G. 1999. Pragmatism, Adaptive Management, and Sustainability. *Environmental Values,* general, 451–466.

Norton, B. G. 2003a. Ecological Integrity and Social Values: At What Scale? In: Norton, B. G. (ed.) *Searching for Sustainability: Interdisciplinary Essays in the Philosophy of Conservation Biology.* New York: Cambridge University Press.

Norton, B. G. 2003b. What Do We Owe the Future? How Should We Decide? In: Norton, B. G. (ed.) *Searching for Sustainability: Interdisciplinary Essays in the Philosophy of Conservation Biology.* New York: Cambridge University Press.

Norton, B. G. 2005. *Sustainability: a Philosophy of Adaptive Ecosystem Management.* Chicago: University of Chicago Press.

Norton, B. G. & Ulanowicz, R. E. 2003. Scale and Biodiversity Policy: A Hierarchial Approach. In: Norton, B. G. (ed.) *Searching for Sustainability: Interdisciplinary Essays in the Philosophy of Conservation Biology.* New York: Cambridge University Press.

O'Connor, M. 1998. Ecological-Economic Sustainability. In: Faucheux, S. & O'Connor, M. (eds) *Valuation for Sustainable Development.* Northampton, MA: Edward Elgar.

O'Riordan, T. 2005. On Justice, Sustainability, and Democracy. *Environment,* 47(6), 0.

O'Rourke, D. & Connolly, S. 2003. Just Oil? The Distribution of Environmental and Social Impacts of Oil Production and Consumption. *Annual Review of Environment and Resources,* 28, 587–617.

Obst, C. 2000. Report of the September 1999 OECD Expert Workshop on the Measurement of Sustainable Development. In: Organisation for Economic Co-operation and Development (ed.) *Frameworks to Measure Sustainable Development: An OECD Expert Workshop*. Paris: OECD.

OED 2011a. Sustain, n. *Oxford English Dictionary*. Oxford: Oxford University Press.

OED 2011b. Sustain, v. *Oxford English Dictionary*. Oxford: Oxford University Press.

Olalla-Tárraga, M. Á. 2006. A Conceptual Framework to Assess Sustainability in Urban Ecological Systems. *International Journal of Sustainable Development & World Ecology*, 13, 1–15.

Opschoor, J. B. 1997. Forum: The Hope, Faith and Love of Neoclassical Environmental Economics. *Ecological Economics*, 22, 281–283.

Ott, K. 1994. Ethical Questions Embedded in Water Resource Provisions of UN Agenda 21. In: Brown, D. (ed.) *Proceedings on Interdisciplinary Conference Held at the United Nations on the Ethical Dimensions of the United Nations Program on Environment and Development, Agenda 21*. New York: Earth Ethics Research Group, Northeast Chapter, United Nations.

Ott, W. R. 1978. *Environmental Indices: Theory and Practice*. Ann Arbor, MI: Ann Arbor Science Publishers.

Ozdemir, I. 2003. Toward an Understanding of Environmental Ethics from a Qur'anic Perspective. In: Foltz, R. C., Denny, F. M. & Bahruddin, A. (eds) *Islam and Ecology: A Bestowed Trust*. Cambridge, MA: Harvard University Press.

Paden, R. 1994. Free Trade and Sustainable Development – The Moral Basis of Agenda 21 and Its Problems. In: Brown, D. (ed.) *Proceedings on Interdisciplinary Conference Held at the United Nations on the Ethical Dimensions of the United Nations Program on Environment and Development, Agenda 21*. New York: Earth Ethics Research Group, Northeast Chapter, United Nations.

Palmer, J. A. 1992. Towards a Sustainable Future. In: Cooper, D. E. & Palmer, J. A. (eds) *Environment in Question: Ethics and Global Issues*. New York: Routledge.

Pane Haden, S. S., Oyler, J. D. & Humphreys, J. H. 2009. Historical, Practical, and Theoretical Perspectives on Green Management: An Exploratory Analysis. *Management Decision*, 47, 1041–1055.

Paracchini, M. L., Pacini, C., Calvo, S. & Vogt, J. 2008. Weighting and Aggregation of Indicators for Sustainability Impact Assessment in the SENSOR Context. In: Heming, K., Pérez-soba, M. & Tabbush, P. (eds) *Sustainability Impact Assessment of Land Use Changes*. New York: Springer-Verlag.

Paracchini, M. L., Pacini, C., Jones, M. L. M. & Pérez-Soba, M. 2011. An Aggregation Framework to Link Indicators Associated with Multifunctional Land Use to the Stakeholder Evaluation of Policy Options. *Ecological Indicators*, 11, 71–80.

Parfit, D. 2010. Energy Policy and the Further Future: The Identity Problem. In: Gardiner, S. M., Caney, S., Jamieson, D. & Shue, H. (eds) *Climate Ethics: Essential Readings*. Oxford: Oxford University Press.

Pearce, D. 1997. Forum: Substitution and Sustainability: Some Reflections on Georgescu-Roegen. *Ecological Economics*, 22, 295–297.

Pearsall, H. 2010. From Brown to Green? Assessing Social Vulnerability to Environmental Gentrification in New York City. *Environment and Planning C-Government and Policy*, 28, 872–886.

Peeples, J. A. & Deluca, K. M. 2006. The Truth of the Matter: Motherhood, Community and Environmental Justice. *Women's Studies in Communication*, 29, 59.

Peet, J. 1997. Forum: "Georgescu-Roegen versus Solow/Stiglitz" ... But What is the Real Question? *Ecological Economics*, 22, 293–294.

Peet, J. & Bossel, H. 2000. An Ethics-Based Systems Approach to Indicators of Sustainable Development. *International Journal of Sustainable Development*, 3, 221–238.

Peirce, C. S. 1931a. Consequences of Common-Sensism. In: Hartshorne, C. & Weiss, P. (eds) *Collected Papers of Charles Sanders Peirce*. Cambridge, MA: Harvard University Press.

Peirce, C. S. 1931b. Consequences of Four Incapacities. In: Hartshorne, C. & Weiss, P. (eds) *Collected Papers of Charles Sanders Peirce*. Cambridge, MA: Harvard University Press.

Peirce, C. S. 1931c. Issues of Pragmatism. In: Hartshorne, C. & Weiss, P. (eds) *Collected Papers of Charles Sanders Peirce*. Cambridge, MA: Harvard University Press.

Peirce, C. S. 1931d. Pragmatism and Abduction. In: Hartshorne, C. & Weiss, P. (eds) *Collected Papers of Charles Sanders Peirce*. Cambridge, MA: Harvard University Press.

Peirce, C. S. 1931e. A Survey of Pragmaticism. In: Hartshorne, C. & Weiss, P. (eds) *Collected Papers of Charles Sanders Peirce*. Cambridge, MA: Harvard University Press.

Pereira, W. 1997. *Inhuman Rights: the Western System and Global Human Rights Abuse*. Goa, India: The Other India Bookstore.

PÉRez-Soba, M., Petit, S., Jones, L., Bertrand, N., Briquel, V., Omodei-Zorini, L., Contini, C., Helming, K., Farrington, J. H., Mossello, M. T., Wascher, D., Kienast, F. & De Groot, R. 2008. Land Use Functions – A Multifunctional Approach to Assess the Impact of Land Use Changes on Land Use Sustainability. In: Heming, K., PéRez-Soba, M. & Tabbush, P. (eds) *Sustainability Impact Assessment of Land Use Changes*. New York: Springer-Verlag.

Pezzey, J. 1992. Sustainability: An Interdisciplinary Guide. *Environmental Values*, 1, 321–362.

Praneetvatakul, S., Janekarnkij, P., Potchanasin, C. & Prayoonwong, K. 2001. Assessing the Sustainability of Agriculture a Case of Mae Chaem Catchment, Northern Thailand. *Environment International*, 27, 103–109.

Prescott-Allen, R. 2001. *The Wellbeing of Nations: a Country-by-Country Index of Quality of Life and the Environment*. Washington, D.C.: Island Press.

Preston, R. 1986. Middle Axiom. In: Childress, J. S. & Macquarrie, J. (eds) *Westminster Dictionary of Christian Ethics*. Philadelphia: Westminter Press.

Rametsteiner, E., Puelzl, H., Alkan-Olsson, J. & Frederiksen, P. 2011. Sustainability Indicator Development – Science or Political Negotiation? *Ecological Indicators*, 11, 61–70.

Redclift, M. R. 1987. *Sustainable Development: Exploring the Contradictions*. New York, Methuen.

Redclift, M. 1993. Sustainable Development: Needs, Values, Rights. *Environmental Values*, 2, 3–20.

Reddy, A. K. N. 2002. Energy Technologies and Policies for Rural Development. In: Johansson, T. B. & Goldemberg, J. (eds) *Energy for Sustainable Development: A Policy Agenda*. New York: UN Development Programme.

Reed, M. S., Dougill, A. J. & Baker, T. R. 2008. Participatory Indicator Development: What can Ecologists and Local Communities Learn from Each Other? *Ecological Applications*, 18, 1253–1269.

Reed, M. S., Fraser, E. D. G. & Dougill, A. J. 2006. An Adaptive Learning Process for Developing and Applying Sustainability Indicators with Local Communities. *Ecological Economics*, 59, 406–418.

Rees, W. E. & Wackernagel, M. 1998. *Our Ecological Footprint: Reducing Human Imact on the Earth*. Gabriola Island, BC: New Society Publishers.

Ricketts, G. M. 2010. The Roots of Sustainability. *Academic Questions*, 23, 20–53.

Robertson, H. G. 2008. Controlling Facilities. In: Gerrard, M. B. & Foster, S. R. (eds) *The Law of Environmental Justice: Theories and Procedures to Address Disproportionate Risks*. 2nd ed. Chicago, IL: American Bar Association Section of Environment, Energy, and Resources.

Robinson, N., With Hassan, P., Burhenne-Guilmin, F. & under the auspices of the Commission on Environmental Law of the World Conservation Union – the International Union for the Conservation of Nature and Natural Resources (eds) 1993. *Agenda 21 & the UNCED Proceedings*. New York: Oceana Publications.

Rolston III, H. 1994. Environmental Protection and an Equitable International Order: Ethics after the Earth Summit. In: Brown, D. (ed.) *Proceedings on Interdisciplinary Conference Held at the United Nations on the Ethical Dimensions of the United Nations Program on Environment and Development, Agenda 21*. New York: Earth Ethics Research Group, Northeast Chapter, United Nations.

Ryall, A. 2007. Access to Justice and the EIA Directive: the Implications of the Aarhus Convention. In: Holder, J. & McGillivray, D. (eds) *Taking Stock of Environmental Assessment: Law, Policy, and Practice*. New York: Routlege-Cavendish.

Rydin, Y. 1999. Can We Talk Ourselves into Sustainability? The Role of Discourse in the Environmental Policy Process. *Environmental Values*, 8, 467–484.

Sagoff, M. 1994. Biodiversity and Agenda 21: Ethical Considerations. In: Brown, D. (ed.) *Proceedings on Interdisciplinary Conference Held at the United Nations on the Ethical Dimensions of the United Nations Program on Environment and Development, Agenda 21*. New York: Earth Ethics Research Group, Northeast Chapter, United Nations.

Sarkar, A. 2009. Role of Inequality and Inequity in the Occurrence and Consequences of Chronic Arsenicosis in India and Policy Implications. *Environmental Justice*, 2, 147–152.

Satterfield, T. A., Mertz, C. K. & Slovic, P. 2004. Discrimination, Vulnerability, and Justice in the Face of Risk. *Risk Analysis: An International Journal*, 24, 115–129.

Schacht, J. 1982. *An Introduction to Islamic Law*. New York: Oxford University Press.

Schlossberg, M. & Zimmerman, A. 2003. Developing Statewide Indices of Environmental, Economic, and Social Sustainability: a Look at Oregon and the Oregon Benchmarks. *Local Environment*, 8, 641–660.

Sessions, G. (ed.) 1995. *Deep Ecology for the Twenty-First Century*. Boston: Shambhala.

Shiva, V. 1992. Recovering the Real Meaning of Sustainability. In: Cooper, D. E. & Palmer, J. A. (eds) *The Environment in Question: Ethics and Global Issues*. New York: Routledge.

Shrader-Frechette, K. (ed.) 1991. *Nuclear Energy and Ethics*. Geneva: WCC Publications.

Shrader-Frechette, K. 2002. *Environmental Justice: Creating Equality, Reclaiming Democracy*. New York, Oxford University Press.

Shrivastava, P. 1995. The Role of Corporations in Achieving Ecological Sustainability. *Academy of Management Review*, 20, 936–960.

Shue, H. 2010a. Deadly Delays, Saving Opportunities: Creating a More Dangerous World? In: Gardiner, S. M., Caney, S., Jamieson, D. & Shue, H. (eds) *Climate Ethics: Essential Readings*. Oxford: Oxford University Press.

Shue, H. 2010b. Global Environment and International Inequality. In: Gardiner, S. M., Caney, S., Jamieson, D. & Shue, H. (eds) *Climate Ethics: Essential Readings*. Oxford: Oxford University Press.

References

Simon, J. L. 1996. *The Ultimate Resource 2*. Princeton, NJ, Princeton University Press.
Smil, V. 2003. *Energy at the Crossroads: Global Perspectives and Uncertainties*. Cambridge, MA: MIT Press.
Solow, R. M. 1997. Reply: Georgescu-Roegen versus Solow/Stiglitz. *Ecological Economics*, 22, 267–268.
Spellerberg, I., Fogel, D., Fredericks, S. E., Butler Harrington, L. M., Pronto, M. & Wouters, P. 2012. Measurements, Indicators, and Research Methods for Sustainability. *Berkshire Encyclopedia of Sustainability*. Great Barrington, MA: Berkshire Publishing Co.
Spilanis, I., Kizos, T., Koulouri, M., Kondyli, J., Vakoufaris, H. & Gatsis, I. 2009. Monitoring Sustainability in Insular Areas. *Ecological Indicators*, 9, 179–187.
Stern, D. I. 1997. Limits to Substitution and Irreversibility in Production and Consumption: A Neoclassical Interpretation of Ecological Economics. *Ecological Economics*, 21, 197–215.
Stern, D. I. & Cleveland, C. J. 2004. Energy and Economic Growth. *Rensselaer Working Papers in Economics*. Troy, NY: Department of Economics, Rensselaer Polytechnic Institute.
Stern, S. N. 2006. *Stern Review: The Economics of Climate Change: Executive Summary (Long)*. London: HM Treasury. Available: http://www.hm-treasury.gov.uk/independent_reviews/stern_review_economics_climate_change/sternreview_index.cfm [Accessed March 7, 2013].
Stiglitz, J. E. 1997. Reply: Georgescu-Roegen versus Solow/Stiglitz. *Ecological Economics*, 22, 269–270.
Stirling, A. 1994. Diversity and Ignorance in Electricity Supply Investment: Addressing the Solution Rather than the Problem. *Energy Policy*, 22, 195–216.
Sylvan, R. & Bennett, D. 1994. *The Greening of Ethics*. Cambridge, UK: White Horse Press.
Taquino, M., Parisi, D. & Gill, D. A. 2002. Units of Analysis and the Environmental Justice Hypothesis: The Case of Industrial Hog Farms. *Social Science Quarterly (Blackwell Publishing Limited)*, 83, 298–316.
Taylor, B. 1991. The Religion and Politics of Earth First! *The Ecologist*, 21, 258–266.
Taylor, B. R. 1995. Earth First! and Global Narratives of Popular Ecological Resistance. In: Taylor, B. R. (ed.) *Ecological Resistance Movements: The Global Emergence of Radical and Popular Environmentalism*. Albany: State University of New York Press.
Tester, J. W. 2005. *Sustainable Energy: Choosing Among Energy Options*. Cambridge, MA: MIT Press.
The Earth Charter Commission. 2000. *The Earth Charter*. Available: http://www.earthcharter.org/ [Accessed March 7, 2013].
Thompson, P. B. 2010. *The Agrarian Vision: Sustainability and Environmental Ethics*. Lexington: The University Press of Kentucky.
Tisdell, C. 1997. Forum: Capital/Natural Resource Substitution: the Debate of Georgescu-Roegen (through Daly) with Solow/Stiglitz. *Ecological Economics*, 22, 289–291.
Tschirley, J. B. 1997. Box 3K: The Use of Indicators in Sustainable Agriculture and Rural Development: Considerations for Developing Countries. In: Moldan, B., Billharz, S. & Matravers, R. (eds) *Sustainability Indicators: a Report on the Project on Indicators of Sustainable Development*. New York: John Wiley & Sons.
Tucker, M. E. 1994. The Role of Religion in Forming an Environmental Ethics. In: Brown, D. (ed.) *Proceedings on Interdisciplinary Conference Held at the United Nations on the Ethical Dimensions of the United Nations Program on Environment and*

Development, Agenda 21. New York: Earth Ethics Research Group, Northeast Chapter, United Nations.

Tucker, M. E. 2004. Scanning the Horizon for Hope. *Conservation Biology*, 18, 299–300.

United Church of Christ Justice and Witness Ministries. 2007. *Toxic Wastes and Race at Twenty: 1987–2007.* Cleveland: United Church of Christ.

United Nations. 2001. *Indicators of Sustainable Development: Guidelines and Methodologies.* New York: United Nations.

United Nations. 2012a. *The Future We Want.* Available: http://www.uncsd2012.org/content/documents/727The%20Future%20We%20Want%2019%20June%201230pm.pdf [Accessed March 18, 2013].

United Nations. 2012b. *Rio+20 United Nations Conference on Sustainble Development: Objectives and Themes.* United Nations. Available: http://www.uncsd2012.org/objectiveandthemes.html [Accessed March 7, 2013].

United Nations Conference on Environment and Development. 1992. Rio Declaration on Environment and Development. United Nations. Available: http://www.un.org/documents/ga/conf151/aconf15126-1annex1.htm [Accessed March 18, 2013].

United Nations Conference on The Human Environment. 1972. Declaration of the United Nations Conference on the Human Environment. Stockholm: UN.

United Nations Development Programme. 2005. *Energizing the Millennium Development Goals.* New York: United Nations Development Programme.

US Environmental Protection Agency. 2006. *How is the UV Index Calculated?* US Environmental Protection Agency. Available: http://www.epa.gov/sunwise/uvcalc.html [Accessed March 7, 2013].

Van Den Begh, J. C. J. M. & Verbruggen, H. 1999. Spatial Sustainability, Trade and Indicators: An Evaluation of the "Ecological Footprint". *Ecological Economics*, 29, 61–72.

Van Dieren, W. (ed.) 1995. *Taking Nature into Account: a Report to the Club of Rome: Toward a Sustainable National Income.* New York: Copernicus.

Van Wensveen, L. 2000. *Dirty Virtues: The Emergence of Ecological Virtue Ethics.* New York: Humanity Books.

Vatn, A. 2005. Rationality, Institutions and Environmental Policy. *Ecological Economics*, 55, 203–217.

Verburg, R. M. & Wiegel, V. 1997. On the Compatibility of Sustainability and Economic Growth. *Environmental Ethics*, 19, 247–265.

Walter, C. & Stuetzel, H. 2009. A New Method for Assessing the Sustainability of Landuse Systems (I): Identifying the Relevant Issues. *Ecological Economics*, 68, 1275–1287.

Waltzer, M. 1994. *Thick and Thin: Moral Argument at Home and Abroad.* London: University of Notre Dame Press.

Warren, K. J. 1994. Ecofeminism and Agenda 21: a Philosophical View on Taking Empirical Data Seriously. In: Brown, D. (ed.) *Proceedings on Interdisciplinary Conference Held at the United Nations on the Ethical Dimensions of the United Nations Program on Environment and Development, Agenda 21.* New York: Earth Ethics Research Group, Northeast Chapter, United Nations.

Waskow, A. 2003. And the Earth is Filled with the Breath of Life. In: Foltz, R. C. (ed.) *Worldviews, Religion, and the Environment.* Belmont, CA: Thompson Wadsworth.

Wei, Y. P., White, R. E., Chen, D., Davidson, B. A. & Zhang, J. B. 2007. Farmers' Perception of Sustainability for Crop Production on the North China Plain. *Journal of Sustainable Agriculture*, 30, 129–147.

Weida, L. 2009. Climate Change Policies in Singapore: Whose "Environments" Are We Talking About? *Environmental Justice*, 2, 79–83.

Weiming, T. 2001. The Ecological Turn in New Confucian Humanism: Implications for China and the World. *Daedalus*, 130, 243–264.

Weiss, B. G. 1998. *The Spirit of Islamic Law*. Athens: University of Georgia Press.

Weiss, S. D. 1994. Ethical Issues in Toxic Waste Export. In: Brown, D. (ed.) *Proceedings on Interdisciplinary Conference Held at the United Nations on the Ethical Dimensions of the United Nations Program on Environment and Development, Agenda 21*. New York: Earth Ethics Research Group, Northeast Chapter, United Nations.

West, T. C. 2006. *Disruptive Christian Ethics: When Racism and Women's Lives Matter*. Louisville, KY: Westminster/John Knox Press.

White, R. A. 1994. A Baha'i Perspective on an Ecologially Sustainable Society. In: Tucker, M. E. & Grim, J. A. (eds) *Worldviews and Ecology: Religion, Philosophy and the Environment*. Maryknoll, NY: Orbis.

Williams, B. L. & Florez, Y. 2002. Do Mexican Americans Perceive Environmental Issues Differently Than Caucasians: A Study of Cross-Ethnic Variation in Perceptions Related to Water in Tucson. *Environmental Health Perspectives*, 110, 303–310.

Wilson, S. M., Heaney, C. D., Cooper, J. & Wilson, O. 2008. Built Environment Issues in Unserved and Underserved African-American Neighborhoods in North Carolina. *Environmental Justice*, 1, 63–72.

Wogaman, J. P. 1993. *Christian Ethics: A Historical Introduction*. Louisville, KY: Westminster/John Knox Press.

Women in Europe for a Common Future. 2012. *Rio+20: Women "Disappointed and Outraged"*. Women in Europe for a Common Future. Available: http://www.wecf.eu/english/press/releases/2012/06/womenstatement-outcomesRio.php [Accessed March 7, 2013].

Woodruff, T. J., Parker, J. D., Kyle, A. D. & Schoendorf, K. C. 2003. Disparities in Exposure to Air Pollution during Pregnancy. *Environmental Health Perspectives*, 111, 942–946.

World Commission on Environment and Development. 1987. *Our Common Future: Report of the World Commission on Environment and Development*. New York: Oxford University Press.

World Council of Churches. 2013. About the WCC. Geneva: World Council of Churches. Available: http://www.wcc-coe.org/wcc/english.html [Accessed March 7, 2013].

World Council of Churches, Central Committee. 1975. *Uppsala to Nairobi, 1968–1975: Report of the Central Committee to the 5th Assembly of the World Council of Churches / general editor: David Enderton Johnson*. New York: Published in collaboration with the World Council of Churches by Friendship Press.

World Council of Churches, Central Committee. 1983. *Nairobi to Vancouver, 1975–1983: Report of the Central Committee to the Sixth Assembly of the World Council of Churches*. Geneva: World Council of Churches.

World Council of Churches, Central Committee. 1990. *Vancouver to Canberra, 1983–1990: Report of the Central Committee of the World Council of Churches to the Seventh Assembly / edited by Thomas F. Best*. Geneva: WCC Publications.

Yale Center for Environmental Law & Policy, Center For International Earth Science Information Network. 2010. *Environmental Performance Index 2010*. Available: http://www.epi2010.yale.edu/ [Accessed March 18, 2013].

Yale Center For Environmental Law & Policy, Center For International Earth Science Information Network: Emerson, J., Levy, M., Esty, D. C., Mara, V., Kim, C., De Sherbinin, A. & Srebotnjak, T. 2010. *2010 Environmental Performance Index*. New Haven, CT: Yale University Press.

Index

accessibility of indexes 58, 63, 66, 71, 113; definition 65; for policy-makers 159–60, 184–87, 188–90, 193; *see also* indicator, benefits and weaknesses in comparison to indexes; data limits; polarities of index development

activism 18, 37, 45, 47, 147

actual data for indexes 58, 63–66, 71, 110, 113, 181; examples of 73, 134–38; *see* ideal data; adequate assessment of the situation; data limits; ideal data for indexes; polarities of index development

adaptability 7, 85, 96, 112 195–96; in Agenda 21 12, 33–34, 36, 84; as a broad principle 90, 104–5, 113–14; in carbon emissions indexes 126; of comparison 84; definition of 113–14; in the Earth Charter 117; in the Environmental Performance Index 157, 159; of ethical methods 118–21; in the Eurostat's Sustainable Development Indicators 160–61; in *The Future We Want* 52; in indexes in general 13, 169 187; in local indexes 166; in *Our Common Future* 22–23; in the three-dimensional index of SED 129–30, 132–34, 136–39, 143–48; in the Wellbeing of Nations 153–54

Agenda for the Twenty-First Century (Agenda 21) 12, 16, 21, 28–29, 118; adaptability 104–5; adequate data assessment 102–4; basis for comparative ethics 84–85, 88, 109, 112, 121–22; cooperation 108–9; efficiency 105–8; equity 100–102; ethical critiques of 29, 35–36; explicit ethics in 30–31; farsightedness 98–100; feasible idealism 109–12; history of 28; implications of 37, 39–40, 50, 164; justice 12, 84, 85; relationship to *The Future We Want* 52–55; relationship to Rio Declaration 28–29; responsibility 96–98; as a source of ethics 31–36, 72;

agriculture 26, 80; environmental pressure 125, 149, 176; farsighted 98; indicators of 152, 155, 157, 161, 165–66; Islamic 111; sustainable 16, 37–38; a sustainability component 19, 21, 31, 35–37, 51–54, 60

anthropocentrism 43; Agenda 21 36, 45; in broad principles 74, 113–14; Earth Charter 29–30; FAO 38; Nash 97; Rio Declaration 29–30; three-dimensional index of SED 128–29; WCC 13;

adequate assessment of the situation 7, 12–13, 112; in Agenda 21 31–33, 84; as a broad principle 85, 96, 102–4, 112–13; in carbon emissions indexes 124–26; definition of 113; in the Earth Charter 117; in the Environmental Performance Index 156; of ethical methods 119, 121; ethical use of 122, 169, 171, 174, 195–96; in the Eurostat's Sustainable Development Indicators 159–61; in *The Future We Want* 53; in local indexes 165–66; in *Our Common Future* 22; in the three-dimensional index of SED 130–32, 134–36, 140–42, 143–45; in the Wellbeing of Nations 148–52

asthma 47, 172, 176–77, 189

atmosphere 28, 59, 62, 124–26

Bennett, David 72, 85; worldview 87–88, 93–96; adaptability 105;

220 Index

adequate data assessment 104; conserving 107; cooperation 108–9; equity 101; farsightedness 100; feasible idealism 111; responsibility 98
bible 86, 89, 118
biocentrism 36, 74, 94, 111, 114, 118
biodiversity 34, 44, 52, 71; Nash 86, 97, 101; indicators of 155, 161, 163
biomass 131–32, 138–39
bioresponsibility 86, 97
biotic rights 90, 95–97, 101–2
Bossel, Harmet 5, 45, 122
Bowersox, Joe 45
Brandt Commission 18–19, 21
Bretton Woods Institutions 17
broad principle 8, 11, 47, 72–74; definition 82–84; comparison to yield 85, 88, 95, 96–111; definition 82–84; implementation *see* index, ethics in; list of 112–16; method to use 118–21, 191–98; response to critics 73–74, 116–18; specification of 36–37, 101–8, 121, 126, 129, 133, 147; *see also* adequate assessment of the situation, adaptability, careful use, cooperation, equity, farsightedness, feasible idealism, justice, responsibility, signs
Brundtland Report (*Our Common Future*) 4, 7, 9, 15; critics of 23–24, 43; definition of sustainability 4, 21–23, 32, 50; history 21–23; implications 25, 28, 37; justice 7, 50;
Bullard, Robert 5, 47, 132, 175–76; critiques of Agenda 21 36, 101
Business index 42

capitalism 17, 29, 42, 80
carbon dioxide 62–64, 123–24, 130, 167
carbon emissions 21, 62; environmental harm 20, 63; policy 20, 22, 173; justice 173
carbon emissions indicators 1, 12, 62, 123–24; in the Environmental Performance Index 157; ethics of 124–27, 159; local 167; three-dimensional index of SED 130; types of 124–26
careful use 7, 12, 112; in Agenda 21 12; as a broad principle 85, 96, 105–8, 112, 115; in carbon emissions indexes 126–27; definition of 115; in the Earth Charter 116; in the Environmental Performance Index 155–56; of ethical methods 119–21; ethical use of 122, 169, 195–96; in the Eurostat's Sustainable Development Indicators 161, 164; in *The Future We Want* 52; in local indexes 167; in *Our Common Future* 22–23; in the three-dimensional index of SED 127, 133–34, 142–44, 146–47; in the Wellbeing of Nations 149;
CFCs 20
Chavez, Cesar 47, 176
Christianity 12, 72, 93, 118; contributions to broad principles 96–112; *For the Common* Good 27; Nash, James A. 85–88, 90, 106, 109; in sustainability history 16–18, 27; Muelder, Walter 79–81
coal 17, 39, 70, 125, 145
Cobb, John B. Jr. 5–6, 9, 16; economics 24–27, 45, 141; ethics 24–27, 56; sustainable development 24–27;
Comparative Religious Ideas Project (CRIP) 78, 79
comprehensiveness of indexes 10, 58, 60–63, 66; of human wellbeing 41, 177–78; in indexes 28, 64, 66, 71, 113, 120, 123, 141, 154–56, 160; of justice 13, 174, 177–78; of security 144; *see also* adequate assessment of the situation; indicator, benefits and weaknesses in comparison to indexes; and polarities of index development
Confucianism 78; *see also* religion
conserving 107
consumers 3, 41, 44, 71, 84, 162
consumption effects of 18, 86; limiting 23, 29, 35, 105–7, 161; energy 62; Agenda 21 28–30, 32, 34, 105–7; responsible 80, 83; over- 106–7; three-dimensional index of SED 130, 133–67, 143–47; Wellbeing of Nations 145, 150–51; Sustainable Development Indicators 161
cooperation 8, 15, 45, 79, 112; in Agenda 21 31, 34, 36, 84–85; as a broad principle 85, 108–9 113, 116; definition of 113, 116; in *The Future We Want* 52
Copenhagen 173

Daly, Herman 122; *see also* Cobb, John B. Jr.
data limits 134, 153, 164–65, 182–84; *see also* ideal data for indexes

Index 221

deep ecology 12, 72, 121; in broad principle development 12, 72, 85, 87–88; adaptability 150; adequate assessment of the situation 104; biospheric egalitarianism 87, 94–95; careful use 107; cooperation 108–9; EarthFirst 94; farsightedness 100; feasible idealism 111; holism 95; intrinsic value 87, 93, 118, 196; perspective on science 87–88; responsibility 98; vs. shallow 103; worldview 85, 87–88; *see also* Naess, Arne; Bennett, David

deforestation 35, 61

democracy 117, 163–64

developed countries 32, 62; aid other nations 34, 35; energy 124; emissions 126; goals for 33; data prevalence 66; index prioritizes 132, 139, 142, 144, 152, 157, 196; justice 35; wellbeing 93, 126, 137

developing countries: in *Agenda 21* 112; assistance to 112; energy 134–36; emissions 126; index challenges 132, 142; 144, 152, 157, 196; *see also* justice

disarmament 19–21

disease 30, 156–58, 169, 175–76, 189, 198

distributive injustice 48–49, 89, 124, 172; *see also* distributive justice, justice

distributive justice 13, 48–49, 89, 124; goods and services distributed 48, 158, 163; indicators 158–59, 163; methods of monitoring 13, 174, 188–89; relationship to participatory justice 49; who experiences 47–48, 54, 163, 176, 178–79; *see also* distributive injustice, justice

Driving-Force–State–Response (DSR) Framework 10, 39, 62–63, 130

drought 53, 126

Earth Charter 11–12, 29–30, 84; history 28–29; limits 73; relation to Rio Declaration 28–30, 36

Earth Summit 9, 15–16, 184; documents 27–30; influence 51, 128, 184; *see also* Agenda 21, Rio Declaration

Ecological Footprint 40

ecology 8; Brundtland Report 21, 24; division from deep ecology 103; use with ethics 86; 120; use in indexes 39–41, 166; *see also* adequate assessment

Ecosystem Wellbeing Index (EWI) 148–53, 169

efficiency 3, 112; Agenda 21 31, 34, 36, 84; as a broad principle 85, 105–8, 125; carbon indicators for 125–26; energy 41, 126, 128; indicators 3; *Our Common Future* 23; indicators in the three-dimensional index of SED 140–43, 146

egalitarianism 87, 93–95, 101

EJView 191–92

Electricity: access 134; affordability 135; emissions from generation 125, 157; indicators 134–37, 147, 157–58, 171; relation to quality of life 134–36;

Elling, Bo 6, 45, 50, 198

energy 3, 5; efficiency 3, 84, 123; indicators of 2–3, 6, 11, 148–52, 154–55, 161–63, 165, 167–68; *see also* carbon emissions indicators, three-dimensional index of SED; policy 1112; quality of life 123, 134–38, 167

environmental injustice; *see* environmental justice

environmental justice 3, 10, 46–50; definition 47–48; maps 188–89; methods of monitoring 23, 132, 174–79, 184–87; sustainability relationship 50, 171–73, 179–83, 195–99; *see also* distributive justice, justice, participatory justice, restorative justice

Environmental justice movement 10, 46, 47, 50, 51, 172

Environmental Performance Index (EPI) 12, 65, 123, 155–56; ethical analysis of 156–59, 169

Environmental Protection Agency (EPA) 191

environmental racism; *see* environmental justice

environmentalism 19, 29, 86, 99, 126

equity 112; as a broad principle 11, 13, 85–86, 90, 98, 100–102, 105, 112; economic 153, 168; ethical use of 181; in Eurostat's Sustainable Development Indicators 169; household 153; in index theory 5; political 168; *see also* participatory justice; in *Our Common Future* 21; social 153, 168, 189; in sustainability 5; 11, 13, 22, 35–38, 80–81, 100–101,

114, 151, 163, 174; in the *Wellbeing of Nations* 148, 151–54; *see also* justice
equilibrium 25
Eurostat 12, 64, 123, 159–60; *see also* Sustainable Development Indicators, Sustainable Development Strategy
extinction 35, 86, 131, 198

farming; *see* agriculture
farsightedness 7, 12–13, 112; in Agenda 21 31, 36, 84; as a broad principle 85, 96, 98–100, 102, 112, 115; in carbon emissions indexes 126–27; definition of 115; in the Environmental Performance Index 155. 159; of ethical methods 119, 121; ethical use of 169, 171, 174; in the Eurostat's Sustainable Development Indicators 159, 162, 164; in local indexes 167; in *Our Common Future* 22–23; in the three-dimensional index of SED 133–34, 137–39, 143–44, 146–47; in the Wellbeing of Nations 152
feasible idealism 12, 112, 174; in Agenda 21 34; as a broad principle 85, 96, 109–12, 116; in carbon emissions indicators 124–26; definition of 116; in the Earth Charter 117; in the Environmental Performance Index 157 = 8; of ethical methods 119–21; in the Eurostat's Sustainable Development Indicators 122, 169, 196; in local indexes 168, 181, 187–88; of local priorities in indexes 178, 181, 187–88; in the three-dimensional index of SED 132, 138; in the Wellbeing of Nations 154
fishing: Agenda 21 53; contaminated 47, 177; culture and livelihood 176–77; indicators 149, 155, 161–62; knowledge for sustainability 53; safe limits 161; in sustainable development definition 37;
For the Common Good 6, 9, 16, 24, 27, 45
forests in Agenda 21 35; damage to 35, 40, 61; indicators 63, 164–65, 186; knowledge for sustainability 53; in sustainable development definition 37;
fossil fuels: careful use of 116, 128, 144, 161; damage environment 124, 131, 151, 161; diminishing resources 128; Eurostat Sustainable Development Indicators 161; statistics about 132; in

three-dimensional index of SED 132–33, 136, 143–45;
frugality: in Agenda 21 34; index 147; Nash's use of 86, 90, 98, 106–7, 110; subsumed in the broad principle of careful use 106, 108, 123
fuel 126, 128–45, 161–63

gender 48, 54, 115, 153, 163, 189
general 4–5, 9–14, 20–23, 30–57, 61–80, 84, 92, 98, 100, 104, 108, 111, 121–23, 126–27, 138–43, 148, 151, 156, 159, 162, 165, 170, 178, 182–88, 195–98
geographic information system (GIS) 13, 48, 174–79
Gibbs, Lois 47–49, 176
Gillroy, John Martin 45
Global Reporting Initiative 41
God 108, 195; commitment to 74, 107, 108, 118, 120, 205; creator 103; 106; created value 88–89, 97; gift of 101; human role from 80, 96, 99, 100, 107; hope from 110; law of 80, 85–86, 88, 91–93; love 80, 88–89, 96; justice 89; one 90; vague category 76, 78
greenhouse gas; *see* carbon emissions
greenwashing 42
gross domestic product (GDP) 2–4, 165 limits of 64–65, 68, 122, 141, 169; correlation to environmental burden of disease 156; part of an indicator 59–61, 67, 124–25, 140–42, 163
gross national product (GNP) 26, 162

ḥarām 87, 99–100
Hargrove, Eugene 30, 119, 121
herbicide 47, 176
ḥadīth 91, 105–6
ḥimā, 37, 87, 99–101
Hinduism 78; *see also* religion
Human Development Index (HDI) 2, 134–36, 182
Human Wellbeing Index (HWI) 148–49; ethical analysis of 149–53, 169
humility 86, 89–90, 103–4, 108, 113
hydroelectricity 145, 151

ideal data for indexes; *see* actual data; adequate assessment of the situation; data limits; and polarities of index development
index 1–2, 58–60; community defined; *see also* local indexes, participatory justice; ethics in 8–9, 22–23, 66–71,

169–70, 195–99; *see also* adequate assessment of the situation, adaptability, careful use, farsightedness, feasible idealism, justice, responsibility; popularity of 2–3, 39–42, 50, 164; in policy-making 3–4, 39, 44–45, 58–60, 63–65, 67, 71, 74–75, 132, 142, 146, 148–49, 154–56, 158–59, 164, 166–67, 176, 179, 184–86, 189–90, 195; *see also* carbon emissions indicator, Driving-Force-State-Response Framework, Ecosystem Wellbeing Index, Environmental Performance Index, Sustainable Development Indicators, Gross Domestic Product, Gross National Product, Human Development Index, Human Wellbeing Index, Index of Sustainable Economic Welfare, polarities of index development, indicator, local indicators, subindex, subindicator, three-dimensional framework for sustainability indicators, three-dimensional index of sustainable energy development, UV index, *Wellbeing of Nations*
Index of Sustainable Economic Welfare (ISEW) 25–26
indicator 1–2, 58–60; benefits and weaknesses in comparison to indexes 63–64, 68, 125–26, 128, 130, 148–51, 159–60, 181–83, 186, 198; *for examples see* Ecosystem Wellbeing Index, Environmental Performance Index, Sustainable Development Indicators, Human Development Index, Human Wellbeing Index, Index of Sustainable Economic Welfare, local indicators, three-dimensional framework for sustainability indicators, three-dimensional index of sustainable energy development, UV index, *Wellbeing of Nations*; headline 64, 159–60; *see also* indicator
indigenous people: in Agenda 21 28–29, 35, 45, 54; in Earth Charter 30; in *the Future We Want* 54, 64; intrinsic value of 30, 64; justice for 29, 45, 48, 64, 90, 176;
industrialized nations; *see* developed countries
infrastructure 33, 153, 167, 178, 185

intergenerational justice 17, 98, 115; definition 4; indicators 163; in environmental justice 13, 44, 47, 171, 173; in sustainability; 7, 42, 44, 50; *see also* farsightedness, justice
Intergovernmental Panel on Climate Change (IPCC) 176
International Energy Agency's (IEA) 128, 130, 138, 140, 145
interviews 13, 174, 177–79
intragenerational justice; *see* intergenerational justice
intrinsic value: in broad principles 98, 100–102, 104, 113–14; challenge of quantifying 18; Christian 88; critiques of Agenda 21 36; commitment to 118, 120, 204; critiques of indexes 129, 150, 153, 179; deep ecology 85, 86–88, 93–96, 98, 110, 118, 120, 204; in *The Future We Want* 54; 8, 36, 54, 85–88, 93–104, 113–14, 118, 120, 129, 150, 153, 169, 195
invasive species 47, 176
Islam 12, 72, 195; in Agenda 21 31; in comparison to develop broad principles 99–112; 195; worldview 85–87, 90–93
Islamic law 72, 85–93, 195; in comparison to develop broad principles 99–100, 104–11

Judaism 78, 83
justice 1, 6–7, 12, 112, 175; in Agenda 21 12, 84, 85; biotic 36, 99–100, 105–6, 114, 170; as a broad principle 85, 89–91, 93, 96, 100–102, 106, 111, 115, 120; in carbon emissions indexes 126–27; Christian 89–90; definition of 114; Daly, Herman 26; in the Earth Charter 116; in the Environmental Performance Index 155, 167–69; of ethical methods 119, 121; ethical use of 119–21, 122, 169, 195–96; in the Eurostat's Sustainable Development Indicators 159, 162–64; in *The Future We Want* 52–54; Islamic 91–92; in local indexes 165, 167–68; Muelder, Walter 80–81; neglect in neoclassical economics 61; in *Our Common Future* 22, 36, 47; in sustainability definitions 13, 14, 37; in the three-dimensional index of SED 129, 132–34, 137–39, 140, 143–44, 146–48; trickle-down justice 5, 180, 182–83, 196; in the

Wellbeing of Nations 150, 153–55; World Council of Churches 23, 26, 80; *see also* distributive justice, environmental injustice, environmental justice, equity, intergenerational justice, participatory justice, restorative justice

khilāfa 91, 97, 103

Limits to Growth 18
livestock 154, 165
Llewellyn, Othman Abd-ar-Rahman 72, 85, 121, 195; worldview 86–88, 90–93; contribution to broad principles 97–112
local indexes 46, 123; challenges of incorporating into national indexes 179–83; ethical analysis of 164–69; examples 164–69, 176–78; maps 188–94; numerical aggregation methods 184–88, 192–94;
Lopez-Ridaura, S. 6, 12, 46, 165–66
love 77; in Nash's Christian theology 77, 85–90, 97, 109; in Islamic law 106

manageability of indexes; *see* comprehensiveness
manufacturing 123–25, 132, 142, 162
map 175, 189–92
metaphysical: in adequate assessment 113; advantages of broad principles 83, 97, 112, 118; deep ecology 93, 96; possible obstruction of ethical decision-making 6, 74, 118; law 79
middle axioms 79–82; informs broad principles 92–93
Millennium Development Goals (MDG/MDGs) 118, 155, 182
mining 39, 131, 149
minorities: challenges to ethical methods 8, justice for 36, 188, 191–92; risks to *see* environmental justice
Montreal Protocol 20
Muelder, Walter 11, 75, 78; ethics 79–82; middle axioms 79–80
Muslim; *see* Islam deep ecology 12, 72, broad principle development 85, 87–88; adaptability 150; assessment of the situation pheric egalitarianism 87, eful use 107; cooperation

108–9; EarthFirst! 94; farsightedness 100; feasible idealism 111; holism 95; intrinsic value 87, 93, 118, 196; perspective on science 87–88; responsibility 98; vs shallow 103; worldview 85, 87–88; *see also* Naess, Arne; Bennett, David

Naess, Arne 72, 87, 93–96; *see also* deep ecology; Bennett, David; Sylvan, Richard
Nash, James A. 43, 72, 121, 195; adequate data assessment 102–4; adaptability 104–5; careful use 106; cooperation 108; farsightedness 98; feasible idealism 109; frugality 106; humility 103, 108; justice 89, 101; "politicians moral dilemma" 109; responsibility 97; worldview 86, 88–90; *see also* Christianity
national indexes 40, 57, 123, challenges of 165–68, incorporating local into 185–206; *see also* carbon emissions indicators, the Environmental Performance Index, Sustainable Development Indicators, the three-dimensional index of SED, and the *Wellbeing of Nations*,
National Oceanic and Atmospheric Administration (NOAA) 59
Neville, Robert Cummings 11, 57, 75–83
nongovernmental organizations (NGOs): in Agenda 21 28, 35, 53; Rio+20 51–52, 54, 65; sustainability goals 1–2, 4, 10; sustainability indicators 2, 4, 10, 39, 41
normative 5, 9, 15; barriers to influence on indexes 6–8, 23–27; in the Brundtland Report 21–23; in business 41; in ecological studies 39–40; in environmental justice 46–49; *see also* environmental justice; in Rio+20 51–52, 54–55; in the Rio Earth Summit 28–29; *see also* Agenda 21, Rio Declaration; in government 40; in sustainability definitions 19–20, 25–26, 37–39; in sustainability indexes 66–71; *see also* broad principles, indexes
Norton, Bryan G.: index ethics 5–6, 46, 74; justice 23, 43; participatory indicator identification 50, 67, 198; sustainability 43, 56, 57

nuclear 133, 156, 176; indicators 131–33, 156; power 39, 131–32; waste 1, 44, 48, 62, 68, 131–33, 151, 156

objectivity / objective 6, 67, 160, 164, 193
oceans 28, 51, 149
ontological 77–78, 97, 113
Organisation for Economic Co-operation and Development (OECD) 62, 145
Our Common Future (*see* Brundtland Report) 15, 21, 28, 32

Paracchini, Maria Luisa 184–86
participatory injustice 49–50, 54, 158–59, 163–69, 172–73, 178–79, 192, 197–98; *see also* justice, participatory justice
participatory justice 23, 55, 102, 116, 173–74; barriers to 49; Christian 23, 80, 101, 109–10; critiques of *The Future We Want* 65; in *The Future We Want* 53–55; methods of monitoring 13, 174, 178, 188, 190–92; in index development theory 6, 45, 50, 77, 79, 197–98; in indicators 136, 155, 161, 164, 179, 181; relationship to distributive justice 49; who experiences 30, 47–48, 54, 158, 163, 169, 176, 178–79; *see also* justice
Peet, John 5, 24, 45, 122
Peirce, Charles Saunders 11, 75–77
pesticides 156, 157, 176
place-specificity 185, 187, 193
polarities of index development 63, 71, 113; *see also* accessibility of indexes; actual data for indexes; comprehensiveness of indexes; ideal data for indexes; manageability of indexes; technical rigor of indexes; and indexes, ethics in
pollution 86, 93, 101–3, 149, 176; Agenda 21 25; air 25, 40, 71, 136, 138, 152, 156, 167; effects 52; ethics of 91, 99, 101, 105, 113; Islamic prohibitions of 91, 99; land 71, 137; limits of national indexes 181; monitored by the Environmental Performance index 155, 157–58; monitored by Eurostat's Sustainable Development Indicators 161; monitored by local indexes 167, 187; monitored by the three-dimensional index of SED 126, 130–34; monitored by the Wellbeing of Nations 51; regulations 13, 180; water 13, 25, 40, 71, 99, 126, 151, 197; *see also* carbon emissions, justice,
poor: economics 153; energy use 134–35, 137–39, 144, 147, 162; health 155, 158, 169, 176; intrinsic value of 30; Islamic ethics 91, 108; just treatment 45, 54, 100; needs of 22, 106; sustainability definitions include 22, 26; *see also* environmental justice
poverty192; and environmental degradation 17, 23, 28, 32, 35, 43, 62; justice 35; indicators 161–63; *see also* poor
Prescott-Allen, Robert; *see* Wellbeing of Nations
Preservation: biodiversity 52, 97; cultural 1, 55; land 92, 99–100, 167; in sustainability definitions 14–16, 19, 38;
psychological 90, 96, 113, 177, 179, 183

quantitative data 1, 3, 8, 13, 25–26, 40, 56–59, 69, 80, 146, 174, 187, 197
Qur'an 86, 91–92, 105–6
qiyās 92

race; *see* environmental justice
recycling 1, 5; careful use 107–8, 115; emissions reducing 126, 133; energy 126, 142, 148; index 156, 165, 167;
reductionism 3, 10
relationality 86, 97, 102–4
relativism 11, 93, 121
religion 80. 87; comparative 11, 76–79; leaders 1, 29; pluralist challenges to ethics 6, 8, 27, 37, 40, 44, 46, 52–53, 67, 73–74, 85; *see also* Buddhism; Christianity; Cobb Jr., John B.; Confucianism; Hinduism; Islam; Judaism; Llewellyn, Othman ar-Rahman; Nash, James A.
responsibility 7, 12, 70, 112 174; in Agenda 21 31, 35–36, 84–85; as a broad principle 85, 90–91, 95–96, 98, 112, 115–16; business 41; in carbon emissions indexes 124, 126–27; definition of 115–16; in the Earth Charter; in the Environmental Performance Index 158; of ethical methods 119, 121; ethical use of 122, 171, 195–96; in the Eurostat's Sustainable Development Indicators 159, 164; in *The Future We Want* 52;

in indexes in general 67; in local indexes 167–68; in *Our Common Future* 22–23; restorative justice 173; society 18, 80–81; sustainability definition 38; in the three-dimensional index of SED 125, 127, 133–34, 137, 139, 142–43, 146–47; in the Wellbeing of Nations 146, 150–52
restorative justice 49–50, 172; need for 173; methods of monitoring 174, 179; *see also* justice
rights 95, 101; biotic 89–90, 97, 101–2; civil rights movement 11, 48; of future generations 98; human 29, 40, 89, 108; to natural resources 91, 93; of nations 19, 21, 29; in *Our Common Future* 21–22, 29; stakeholder 50; migrants 52; nature 54; to participate in decision-making 34, 67, 108; women's 91
Rio Declaration 16, 118; comparison to other documents 53–54, 56; development of 27–30; ethical critiques 29, 36; rivers 106, 151, 186

Schrader-Frechette, Kristen 44
semiotics 11, 75, 78, 83; *see also* signs
Shari'a 86–87, 91, 99, 111
signs 75, 76, 77, 83, 131, 199; basis of broad principles; general 76–77; iconic 75; indexical 75; symbols 75–79; vague 33–34, 76–78, 82–84; *see also* broad principles, Comparative Religious Ideas Project; Peirce, Charles Saunders; Neville, Robert Cummings; semiotics; Wildman, Wesley
single-indicator index 63
sociological 6, 26, 58, 71, 113
solar 136
Stockholm 17, 128
subindex 2, 59, 65, 68, 143, 186–87; aggregating 174–79; *for examples see* 132, 136–38, 146, 154, 159–60; *see also* Ecosystem Wellbeing Index, Environmental Performance Index, Human Wellbeing Index, index, indicator, manageability, three-dimensional index of SED, and Wellbeing of Nations
survey 59, 62, 141, 153; *see also* 5, 103–6, 109, 153, 167

surveys 13, 174, 177–79, 182, 190, 197
sustainability: definitions 19–24, 28, 37–38; ethics 2–9, 16–20, 22–24, 42–45, 118–21, 195–99; *see* broad principles; history of 16–19; *see also* Agenda 21, *For the Common Good*, Rio+20; index theory *see* index, indicator; indicators of *see* Ecosystem Wellbeing Index, Environmental Performance Index, Sustainable Development Indicators, Human Development Index, Human Wellbeing Index, Index of Sustainable Economic Welfare, local indicators, three-dimensional framework for sustainability indicators, three-dimensional index of sustainable energy development, UV index, *Wellbeing of Nations*; *see also* normative
Sustainable Development Indicators (SDI) 123, 159–64, 169
Sustainable Development Strategy (SDS) 159, 162
sustainable energy development (SED) 126–30, 133–36, 140–47, 169
sustainable growth 25
Sylvan, Richard 72, 85–88, 93–101, 105–11
symbols 59, 65, 75–76, 189, 191

tawḥīd 90
tax 3, 84
technical; *see* normative
technical rigor of indexes; *see* accessibility
Thompson, Paul 18–22, 44
Three-Dimensional Index of Sustainable Energy Development 12, 39, 146–48, 159, 169; data limits 152; goals of 127–30; economic dimension 139–46; environmental dimension 130–34; social dimension 134–39; for ethical assessment of *see* adequate assessment of the situation, adaptability, careful use, farsightedness, feasible idealism, justice, and responsibility
threshold 68, 135–37, 198
toxics 13; environmental justice regarding waste 23, 47–48, 174–76; indicator 156; maps of waste 188–92
tractability 81–82
trickle-down justice 5, 180, 182–83, 196

Index   227

United Church of Christ (UCC) 47
United Nations (UN) 12, 17, 29
United Nations Conference on Environment and Development (UNCED) 28, 51
United Nations Conference on the Human Environment 17
United Nations Development Programme (UNDP) 132, 138
United Nations Food and Agriculture Organization 37
uranium 130
urban 47, 152, 176, 188, 190
utilitarianism 78
utilitarianism 78
UV index 59

vague (vagueness) 38, 75–82, 112, 117, 182, 196
violence 144

waqf 87, 99–101
waste: ecosystem assimilation of 6, 40, 129, 134; Islamic prohibitions against 91, 93, 96, 107; indicators 149, 152, 156, 161–62, 165; policy 1, 28, 34, 52; *see also* nuclear, toxics
water 1, 18, 44, 99; careful use 34, 52; filtration 26, 52, 93, 99, 165, 176; indicators 131, 148–49, 152–53, 161, 165, 169; Islamic law 99–101, 113, 116; policy 34, 35, 51, 188, 191; quality 2, 55, 158; quality of life 44, 91, 172; sustainability component 1, 44, 60–62, 99; in sustainability definition 36; *see also* pollution. rights
Wellbeing of Nations 60, 123, 148, 159, 194
wetlands 26, 156
Wildman, Wesley 11, 75–83
women 92, 138; contributions to sustainability 29, 53; data limitations 53; in *The Future We Want* 53–55; intrinsic value of 30, political participation 33, 158, 169; rights 91; 28–30, 33–35, 48, 53–55, 92, 100, 138, 168–69
workforce 35, 47, 54, 162
World Commission on Environment and Development (WCED) 4, 21
World Council of Churches (WCC) 18, 19, 26, 80
worldview 1, 100; comparison of 78, 81–82; deep ecology as 87–88; ethical consensus among 82–85, 113, 116, 118, 121, 195–99; ethical diversity 2, 15, 20, 73–75, 78, 113, 116, 118; lived 94; local 72–73; new 93, 96; *see also* religion, deep ecology